The Carbonates of Leadville and the Formation of Coal in Colorado

by Fred Stanton

with an introduction by Kerby Jackson

This work contains material that was originally published in 1879.

Introduction

It has been years since the important publication "Carbonates of Leadville: A Treatise on the Formation of Coal in Colorado" was published. First released in 1879, this work has been unavailable to the mining community since those days, with the exception of expensive original collector's copies and poorly produced digital editions.

It has often been said that "*gold is where you find it*", but even beginning prospectors understand that their chances for finding something of value in the earth or in the streams of the Golden West are dramatically increased by going back to those places where gold and other minerals were once mined by our forerunners. Despite this, much of the contemporary information on local mining history that is currently available is mostly a result of mere local folklore and persistent rumors of major strikes, the details and facts of which, have long been distorted. Long gone are the old timers and with them, the days of first hand knowledge of the mines of the area and how they operated. Also long gone are most of their notes, their assay reports, their mine maps and personal scrapbooks, along with most of the surveys and reports that were performed for them by private and government geologists. Even published books such as this one are often retired to the local landfill or backyard burn pile by the descendents of those old timers and disappear at an alarming rate. Despite the fact that we live in the so-called "Information Age" where information is supposedly only the push of a button on a keyboard away, true insight into mining properties remains illusive and hard to come by, even to those of us who seek out this sort of information as if our lives depend upon it. Without this type of information readily available to the average independent miner, there is little hope that our metal mining industry will ever recover.

This important volume and others like it, are being presented in their entirety again, in the hope that the average prospector will no longer stumble through the overgrown hills and the tailing strewn creeks without being well informed enough to have a chance to succeed at his ventures.

Please note that at times it is necessary to rearrange illustration plates in these texts. Any illustrations not found in their original sequence may be found following the index. Regrettably, at this time, we cannot reproduce the large fold out maps that sometimes appeared in the original edition.

Kerby Jackson
June 2016

www.goldminingbooks.com

INTRODUCTION.

It needs no apology for the publication of a Treatise on these subjects. Scarcely anything has ever been written on the metallic carbonates, and therefore this little book, unpretentious as it is, will be hailed with pleasure by the miners of the carbonates. Metallurgical and chemical works have always been written in such abstruse language that the people could not read them. I have aimed to give a common sense, clear and concise account of the carbonates of Leadville, adopting brevity and plainness as the leading objects in view. The recent discoveries of the carbonates in Colorado will, ere long, demand an extensive emigration to the wondrous deposits of the new mining field. This little work may form the key-note for an extensive addition to the literature on the subject; if so, my object is fully answered.

As to my ideas in regard to the origin of the coals of Colorado, they are new, and in submitting them to the criticism of scientists, I do so with diffidence and a desire to elicit truth. In claiming a new discovery of the origin of coal, I hope it will meet with a fair and candid reception at the hands of the scientific world. I have also aimed to simplify assaying as much as possible, so that he who reads may understand. With these few remarks, I submit my little waif to a candid public.

FRED. J. STANTON.

THE CARBONATES OF COLORADO.

As in the ores of Tellurium, and the Tellurides, Colorado comes to the front again in producing and drawing prominently to the attention of the Metallurgical world, a combination of minerals in a deposit which has been hitherto almost unknown. The only place known to produce Tellurium, was the Altai mountains, until Colorado, a few years ago, burst to the view of science with a Tellurium belt unequaled elsewhere, either in the old world or the new.

Now she again surprises the scientist with a metallic combination or association of metals covering such an area. and of such an extent, which is probably alike unparalelled in the literature of science— silver in combination with the carbonates.

In volcanic countries, like the mountains of Colorado, the amount of carbonic acid gas evolved by means of sulphuric acid acting upon, and causing the decomposition of the vegetable accumulation of the vast forests of the Carboniferous period, would cause the formation of carbonates, by the oxidation of the metals in the presence of atmospheric air and carbonic acid. from the same oxides which without such decomposition and the substitution of sulphuric acid, would cause the formation of sulphurets.

It is well known that the percolation of water through porous rocks has an influence upon the character of minerals. Water percolating through

4

or washing over a deposit of carbonate of lime in the form of chalk, marble, calcspar, feldspar, oolite, limestone, shells, etc., again percolating through strata of iron or copper pyrites with silver, would resolve itself into a crystalline deposit of sulphurets, and metallic carbonates would be separated. Thus, carbon has removed the sulphur and the carbonates have reduced and taken the place of the sulphurets. In the reduction of ores, we find that the carbonates and the oxides are much more easily reduced than the sulphurets. Carbon forms a much more important auxiliary in the reduction of the sulphurets than it does of the carbonates and oxides, and the chlorides are more easily reduced than either.

The Colorado ores are chiefly of the order of the sulphurets, and until very lately, miners have had not even the remotest idea, that in the carbonate envelope would be found as rich a kernel of the precious metals as can be met with, either in the free gold or the sulphurets.

The formation of these metallic salts is one of those beautiful processes which is constantly being carried on in the great laboratory of nature. They are chiefly composed of carbonate of lead and antimony, with silver from a very small, to a very large amount; also carbonate of lead, antimony and silver, with copper in the forms of argentiferous, grey copper ore and silver glance. In some cases carrying traces of gold, but as a rule there is no gold.

It is almost beyond dispute, that the carbonates of Leadville and Silver Cliff are merely decompositions of the sulphurets, and that at greater depths, the mineral matter will eventually run into sulphurets. The carbonates, like the chlorides, are what

are known in Europe as "Gossans," a scoriaceous
deposit. Up above the carbonate and chloride belts
will be found at some future time, gold veins in
the decomposed, honey combed, metamorphic rocks,
which may produce very great results, if not found
as Governor Gilpin says: "In Mass and Position,"
possibly in a state of fusion without many peculiarities which have been carried away.

Minerals and Rocks Accompanying the Carbonates.

Auriferous iron, magnetic and copper pyrites, dis-
tributed through the metamorphic and granitic
rocks, which overlie and form the superincumbent
rocks above the carbonates.

Carbonate of copper (malachite and azurite)
formed undoubtedly by the decomposition of the cupreous pyrites, carrying down the metals. except
gold, which is not susceptible of a like oxidation.

Nearly all the carbonaceous rocks are allied and
combined with silver, but no gold.

Galena, native, brittle, horn, dark and light ruby,
glance and fahlera silver ores.

Specular, sparry, red and brown hematite, titanic,
spathic and micaceous iron ores.

Cerussite, anglesite, augite, anhydrite, leucite,
quartz, feldspar, calcspar, gypsum, baryta, crysolite,
obsidian, opal, zinc blende, hornblende.

The interstices between the rocks are filled up by
carbonate of lime, kaolin, baryta, etc,

The presence of large quantities of kaolin (porcelain clay) is caused by the decomposition of the feldspar and the abstraction of the alkalies. It is a white, soft, greasy, soapy-like material and will some day become valuable in the commerce of Colorado.

"A writer in the *Rocky Mountain News*, says: Where porphyry penetrates the mineral bed it is generally soft, sometimes like clay or putty, to be moulded in the hand. There is also common, in or near soft carbonates, masses, or kidneys, of talc, locally known as "Chinese tallow." It is a beautiful lustrous spar, often soft like putty but sometimes quite hard. It breaks or separates with a shining metallic surface; is generally white, but tints into several delicate colors. Animal and vegetable remains—bones, and fragments of trees—are credibly reported to have been found in some of the workings more than a hundred feet below the surface of the ground."

The region of Silver Cliff has a deposite of carbonates, but not as large as the vicinity of Leadville. At the former place, the carbonic acid of the limestones and other sources, has produced carbonates; but there are also large deposits of the chlorides, formed by muriatic or hydrochloric acid, generated by the action of the chloride of sodium (common salt), large quantities of which are found in the neighborhood. Thus, we see sulphuric acid changing the oxides to sulphates; carbonic acid to carbonates, and hydrochloric acid to chlorides.

In volcanic and metamorphic regions, rocks which are chloritic in one locality, may be carbonitic in another, characterized by the feldspar in one and the hornblendes in the other, owing probably to the presence or not of the alkaline materials.

The carbonate veins have evidently originated by means of infiltration from above, filling the seams and fissures.

Dana says:

"Atmospheric waters decompose some species readily (pyrite etc.) and take the new ingredients (sulphate of iron, etc.) into solution. Feldspathic minerals may be decomposed, and the waters thereby become silicious and alkaline. Also, in one way or another, they may become carbonated. Thus armed, the waters go on making various changes in the ores and minerals of the vein, altering chalcopyrite (sulphur of copper and iron) to copper glance or erubesite (sulphids of copper) or to malachite (carbonate of copper), or changing in a similar manner ores of silver or lead, etc. In some parts, the arrangements may be such as to produce a galvanic effect, further promoting decomposition and recomposition. When the solutions differ, after intervals of time, there will be a succession in the changes; and layers of different species may be found."

Thus a layer of quartz, may be succeeded by one of fluorite, or of zinc blende, or calcite, or quartz again, etc. In the course of the changes, a layer of cubes of fluorite, underlying one of quartz, may be entirely dissolved away, and the cubical cavities filled up by another species as blende, etc. The rock of the walls (especially of the lower wall where the vein is inclined) when not united firmly to the vein, often undergoes deep alteration, and may become penetrated by ores from the vein itself, carried in by infiltrating solutions. These alterations are most extensive in the upper part of veins, where it often happens that the metals are removed by infiltrating waters; excepting for the most part the iron, which is left in the red oxide, giving its color to the earthy mass on the top of the vein (called then "the iron hat"). Hence the occurrence of a line of red earth in the soil may be an indication of a valuable mineral beneath.

Gold bearing quartz veins generally lose the pyrite, and perhaps other ores which they contain, and thus become cavernous to a considerable depth. To this distance they are mined with comparative ease ; but beyond, they are extremely hard and much more difficult to work.

It is by the direct infiltration from above that the carbonate of lime is often deposited in superficial seams or open cavities as is so emphatically demonstrated in the region of the carbonates of Colorado.

Professor Charles S. Richardson, G. M. E.— writing to the London *Mining Journal* from Alma, December 23. 1878, describes the geology of the district, and the character of that peculiar metalliferous deposit, as follows :

"The Leadville hills are of Devonian age, stratiform in structure, and reposing uncomfortably on the Silurian below. They consist of trachyte, quartzites, lime and shales. These are both calcareous and siliceous. The lime varies very much in different beds. Some are highly ferruginous, some argillaceous, and others again vitreous or crystalline. The dolomite variety occurs in many places, but it is not granular. Veins of calcite in transparent crystallization are found in the more indurated rocks, associated with the sulphate, and carbonate of magnesia. There are also bed veins of alstonite—barytocalcite, but not in any quanity as far as I have yet seen. There is supposed to be a general contact line between the igneous and aqueous deposits of an uniform structure throughout this locality—the porphyritic rock above, and the limestone below; but I must object positively to this commonly conceived idea, and for one especial reason. I know of many places in these mountains where the opposite is the fact—lime above, and porphyry below.

"Now, so far as the present workings are concerned they are pretty generally confined between the porphyry and the lime ; but one of the largest mines, called the Little Pittsburg, is not covered by any porphyry or trachyte at all. The ore lies on a bed or seam of brown oxide of iron, which is from 1 foot to 20 feet thick.   On this, iron reposes in distorted shapes of various thicknesses and beds of ochreous earths, containing the sulphuret and carbonates of lead. They are of two kinds—the granular and the compact.   The former is in the form of sand or dirt, and can be dug up by the shovel.   It runs from 20 ozs, to 300 ozs. of silver per ton ; the sulphurets of lead average 55 per cent. ; and the hard carbonites from 25 to 45 per cent.   In consequence of the cheap way the mines can be worked, on account of the deposits being so near the surface, and an entire absence of water, they are exceedingly profitable.   In some instances the ore is merely covered over by the debris or mountain wash, which may be called the diluvial of the Devonian era.   These hills are subject to faults and breaks of the formation, and near them the seam of ore is thrown down to depths not as yet ascertained in any part of the district. It is just the same as in the coal measures.   During my recent geological survey I have traced the courses of some of these breaks.   They are not confine l to any regular course. They are the result of the disruption of the formation of subsidences.   They alter the continuity of the seam of ore, making it come to an abrupt stoppage or direct downthrow. In other instances it would appear that while the lime beds were in a semi-plastic state they became squeezed up forming acute wave lines and sometimes folds.   The ore then is in circular pockets, but very rarely ever separating from its iron base.   There must be at this time more than a hundred mines at work, of course, the bulk of them are prospecting operations, but everyone is sanguine af success.   The great aim is to get down through the trachyte porphyry covering. This is of varying thickness ; but in a very few instances, when the contact with the lime has been struck a course of ore has not been found.

The whole belt of the carbonates stretching for many miles in all directions, embracing Leadville Ten-mile, Silver Cliff, California and Horseshoe Gulches to Fairplay and Alma, is almost entirely made up of metamorphic rocks associated with trachyte and porphyry, oxide of iron, carbonate of lime in the form of feldspar and its ever changeful combinations, in every degree and ratio of composition, to the species of kaolin or plastic, soapy porcelain clay.

The characteristic colors of the rocks portray their character.

The red color is due to the oxide of iron from the heating of the limonites.

The yellow is due to the same, differing only in degree of the oxide of iron.

Brown and brownish-black is the predominating color of the rich hard carbonates of Leadville and Silver Cliff, derived from the decomposition of vegetable matter, and the rock will heat white.

If oxide of iron be present, it will heat red.

If oxide of manganese be present it will heat black or bluish-black.

If spotted red, yellow, white and brown, or combinations of those colors, it is owing to the changes of the percolating liquids over different beds, modifying the color of the oxides of iron and manganese.

If green, it is due to the chlorides, sometimes to serpentine, pyroxene or hornblende, and probably copper.

If greenish-black, to hornblende, mica, tourmaline, lapidolite, epidote, etc.

The foliations and fern-like impregnations in these carbonates are due to the oxide of manganese, and rocks containing them are known as landscape or forest rock.

The following table presents a general view of the composition of the more common rock-making materials, showing their similarity:

Silica	Quartz.
Silica, magnesia and water	Talc.
Silica, magnesia and water	Serpentine.
Silica, maguesia and iron	Chrysolite.
Silica, magnesia, lime or protoxide of iron	Pyroxene.
Silica, magnesia lime or protoxide of iron	Hornblende.
Silica, magnesia, alumina and protoxide of iron	Chlorite.
Silica, alumina	Prophyllite.
Silica, alumina	Andalusite.
Silica, alumina	Kaolinite.
Silica, alumina	Cyanite.
Silica, alumina, fluorine	Topaz.
Silica, alumina, oxide of iron	Staurolite.
Silica, alumina, oxide of iron, potash or magnesia	Mica.
Silica, alumina, lime and soda	Scapolite.
Silica, alumina, lime, magnesia, iron or manganese	Garnet.
Silica, alumina, oxide of iron	Epidote.
Silica, alumina, potash	Agalmatolite.
Silica, alumina, potash, soda or lime	Feldspar.
Silica, alumina, alkali, magnesia and boracic acid	Tourmaline.
Silica, alumina, potash	Leucite,
Silica, alumina, soda and potash	Nephelite.

A writer in the Colorado *Tribune*, says:

The so-called carbonate lodes of Leadville are properly and correctly classified as contact veins. They are situated and occur between the lower Silurian limestone, the foot wall, and trapean porphyry, the overlying or head wall. These formations have a strike nearly North and South. and dip or underly nearly due East. The vein matrix consists of peroxide of iron which is the base, carbonate and sulphuret of lead containing chloride and virgin and sulphuret of silver occurring in massive chrystaline condition, intermixed and blended together so generally that it is often difficult to determine which predominates for several feet. A band of quartzose rock sometimes occurs within the line of contact and occasionally blends or mixes up with the metallic ores, often times having the appearance of having been saturated with a solution of silix, and the whole mass thus cemented and consolidated together at the same points. The out crop of the lodes at the surface of the rock in place is hundreds of feet in width or thickness. At other points it pinches up to a mere line of contact. What miners would or do call a lode means that it is the lode or lead to guide the miner, which he follows until the crevice opens up and the matrix comes in again. Then it is called a lode, which means and includes all the matrix or material occurring between the walls. When the metallic compounds make their appearance such as carbonates, chlorides and metallic sulphides, it is then called the ore vein, which means that portion of the lode that is entirely of secondary completion and different from the adjacent walls or country rock and contains minerals of value. It will be observed that owing to the strike and angle of dip of the lode, that the undulations and elevations on the line of the lode cause sharp angles on the line of strike and other local interruptions are the exceptions to the general rule.

Dana says:

"The carbonates have a hardness not exceeding 5., and consequently will not, when pure, strike fire with steel. All effervesce in hot acids, and part of them in cold."

I. CALCITE GROUP—RHOMBOHEDRAL.

Calcite . . . . Lime.
Dolomite . . . . Lime and Magnesia.
Ankerite . . . . Lime, magnesia, manganese and iron.
Magnesite . . . . Magnesia.
Mesitite . . . . Magnesia and iron.
Pistomesite . . . . Magnesia and iron.
Siderite . . . . Iron.
Rhodochrosite . . . . Manganese
Smithsonite . . . . Zinc.

II. ARAGONITE GROUP—ORTHORHOMBIC.

Aragonite . . . . Zinc.
Manganocalcite . . . . Zinc, magnesia and manganese.
Witherite . . . . Barium.
Bromlite . . . . Lime.
Strontianite . . . . Strontia.
Cerussite . . . . Lead.

III. BARYTOCALCITE GROUP—MONOCLINIC.

Barytocalcite . . . . Lime and baryta.

IV. PARISITE GROUP—CARBONATES CONTAINING FLUORINE.

Parisite . . . . Cerium, lanthanum, bismuth, zinc and fluorine.
Kischtimite . . . . Cerium, lanthanum and fluorine.

V. PHOSGENITE GROUP—CARBONATES CONTAINING CHLORINE.

Phosgenite . . . . Lead and chlorine.

The carbonates occupy an important niche in chemistry and mineralogy, and form a very large and important group of salts, some of which occur in nature in large quantities. They are mineralized and formed by the union of carbonic acid with an alkaline, earthy or metallic base, in the same manner as the sulphurets are formed by sulphuric acid. The affinity, however, of the carbonic acid for the majority of the metals, is so slight that the addition of almost any other acid will cause their decomposition, effervescence from the escape of the carbonic acid gas taking place. By this they are known from other salts. Numerous kinds of the carbonates are native productions; some are decomposed by heat, as that of lime; others suffer no change; while there are some which have a part only of their acid driven off by this agent. In the neutral carbonates, the proportion of oxygen in the base is to that in the acid as 1 to 2.

Effervescence on the application of acids is the peculiar property of the carbonates.

Carbonic acid with lime forms carbonate of lime or common limestone, with oxide of iron it forms spathic or sparry iron; wth oxide of zinc it forms calamine, and so on with other metallic bases.

The chemical analysis of the numerous carbonates, alkaline, earthy and metallic, is easily made by estimating (by weight or measure) the quantity of carbonic acid gas evolved through the action upon them of sulphuric, nitric or hydrochloric acid.

Overman says :

"Carbon has only a feeble affinity for metals, and cannot readily be combined with them. But in most cases, the metals, when reduced from porous oxides, in the presence of an excess of carbon, absorb some of it, and condense it in their pores. It is doubtful if a chemical combination is formed ; still there are indications of legitimate compounds under certain conditions. The best means of forming carburets are the carbonates and oxalates, heated in the presence of carbon. The crude iron obtained from the smelting of sparry iron ore, may be considered a real carburet of iron. Carbonate of lead, when reduced by means of carbon, forms also a carburet ; but this is less distinct than that of iron. In consequence of the faint affinity of carbon for the metals, they are generally very brittle when the amount of it is large. But, when a small amount only is mixed mechanically with metal, as is the case with gray cast iron, its strength is not much impaired The combinations of carbon and metal are more fusible than pure metals ; and as carbon is easily removed from metal by oxygen, it is one of the best means to cause metals to be fusible."

Prof. Henrich, in his letter to the Engineering and Mining Journal, says the galena ores of Leadville consist of hard and soft carbonates of lead and streaks of galena :

"As a general rule, the soft, heavy carbonates run highest in silver and the hard carbonates lowest, while the galena seems to range between them in its percentage of silver. Silica occurs in all of these ores. In the galena it occurs as quartz crystals. As a general rule, the galena is lowest in its percentage of silica and highest in lead, while in silver it runs from thirty to a few hundred ounces to the ton. Highest in silica are the soft, sandy carbonates of these ore beds, the silver ore PAR EXCELLENCE of this district. Here again, as in the iron ore, we have sandy carbonates, which are hardly anything more than a porphyry sand—named so, too, by the miners—carrying a small percentage of lead and a few ounces of silver to the ton, up to siliceous carbonates of lead carrying over sixty per cent. of lead and hundreds of ounces of silver to the ton. When carrying a very high percentage of lead, these sandy carbonates are almost always mixed with small pieces of galena, which accounts for their high percentage of lead, besides carrying considrable silica, probably in the form of quartz-sand. left from the decomposition of the porphyry, without improving its contents in silver."

ELVANS.

An elvan is a dyke of porphyry, and is found in granitic formations, also in the slates. They are a felsite and composed of feldspar, quartz, amphibole, mica, tourmaline etc. In clay slate usually composed of feldspar and quartz; in granite, of feldspar and mica. Iron and cupreous pyrites and tin are often associated with them displacing crystals of feldspar. The strike of an elvan is N. E. to S. W. The dip is generally 40 to 60 degrees. They mostly incline more to the North than to the South. No regular veins of mineral are ever found in them, though they often contain mineral in irregular deposites or beds. Elie de Beaumont says, that the elvans are older than the carboniferous formation, and other writers insinuate that their mineral deposites were formed by sublimation; but at Leadville the opposite is most apparent, as it is found at a much lower altitude than the mother rocks and eventually deposited in a semifluid condition.

CARBONATE OF SILVER.

There is a grave doubt in the minds of mineralogists whether there is such a mineral as the Carbonate of Silver; some contending there is such a chemical combination; while others contend it is only mechanical, and that it is free or native silver in a carbonate envelope, as the carbonate of lead.

We are inclined to favor the latter opinion, though we shall give the opinions of mineralogists adversely and allow the reader to form his own opinions.

The carbonate of silver as it is called by some few writers, or the carbonate of the oxide of silver, is composed of oxide of silver, 72.5; carbonic acid, 12.2; oxide of antimony, 15.3; equalling 100 parts.

It has always been considered a rare mineral; it is of a white color which changes upon exposure to the air to a blackish gray metallic appearance. and is very brittle.

It can be formed artificially as a pale, yellowish white insoluble compound, by adding carbonate of potassa to a solution of nitrate of silver, (Lunar caustic). Until very lately, the carbonate of silver was almost unknown and occurred only in the mine of Wenceslas near Wolfbach or the Black Forest in Europe.

A carbonate of silver ore is thus described by Dana under the head of "Selbite "

"SELBITE. Luftsaures Silber (from anal. by Selb) *Widenmann* Min., 689, 1794; *Lens*, Min., 95, 1794; Gransilber; carbonate of silver; *Selb*, Tasch, Min., XI, 394, 1817; SELBIT, *Haid*, handb'k., 506, 1845. A grayish ore, made a carbonate by Selb, its discoverer in 1788, at the mine Wenzel near Wolfach, with the composition (Widenmann, l. c., here cited from Lenz, l. c.), carbonic acid 12, oxide of silver, 72.5, antimony 15.2 with carbonic acid and oxide of copper. According to Wachner, (Mag. of. Pharm., XXV. 1) it is only a mixture; and according to Sandberger (Jahrb. Min. 1864, 221), one of Selb's original specimens, under lens, proved to contain within earthy argentite, besides dolomite and silver, and all parts afforded a sulphur reaction."

Del Rio, described a carbonate of silver from Real Catorce, Mexico, where it is called *Plata Azul* (Gilb , Ann., LXXI. 11). which also is regarded as a mixture.

In Mexico there is an earthy oxide of iron singularly enough called *Colorados*, and in Peru the same occurs by the name of *Pecos.* which carry silver ; these ores occur with the same characteristics as the Leadville *Gossans*, on the surface and have evidently in the same manner undergone partial decomposition and are found closely connected with the Sulphurets.

Van Cotta says; " The Wenzel mine, in the Wolfbach, Black Forest district, gave large dividends for a long succession of years in the 18th century. The Wenzel lode, which descended in a zig-zag form, contained, with a breadth of 6 inches to 2 feet; heavy spar, brown spar, calc spar, fluor spar with galena, copper pyrites, spathic iron, argentiferous tetrahedrite, ruby silver, dyscrasite, silver glance and native silver. The dyscrasite occurred in masses by the hundred weight, and Selb (the discoverer of the so called carbonate of silver,) saw in 1787, a mass of native silver, weighing 75 pounds and surrounded by silver glance and ruby silver."

This evidence seems conclusive, that the silver of Leadville is native silver and assimilates very much with the Wenzel mine.

He says further :

"In the Sophie mine in the Wittich district, Kinzig Valley, about 35 fathoms (210 feet) below the surface, a branch of heavy native silver, was followed in the hanging wall of the otherwise barren lode. After it had been followed for a few inches, the branch wedged out and the formerly white and very firm granite was found changed to a reddish brown and less firm condition, in which threads of native silver could be seen. On this account the work was prosecuted in the reddish brown granite in a neighboring hollow, and discovered after digging for a few fathoms, a broad leader of silver, which continued so long as the granite retained this color and softness."

The attentive reader must draw comparison between the Leadville deposite and those last mentioned, particularly bearing in mind our remarks upon the formation of these deposites and the similarity of their continuation "in a neighboring hollow."

We must repeat what we said before, that these mines of Leadville are native silver and not carbonate of silver mines, and that ere long masses of native silver will be found of immense size, to greet the wondrous view of the miner, in these Gossans and Elvans of Colorado.

There is also a silver ore found at Wolfach, analyzed by Klaproth, with the same characteristics as found at Leadville, and classified by Dana as Dyscrasite, and coarsely granular. It is an Antimonial Silver Ore, with silver, 78; and antimony 22.

Domeyko found a mass of ore from Chanarcillo, Chili, which was mainly impure chloro-bromide of silver externally, to contain within 55.9 per cent of chloride of silver; 15.1 of antimonid of silver, with

14.5 of carbonates and 14.2 of ochreous clay which will apply particularly to the Gallagher mine, Leadville.

The mines of Leadville, Colorado, assimilate very closely to the mines of Reese River, Nevada. We find the ores rich in silver to be those known as the gray copper ore, with the sulphur displaced by carbon. They are classed as antimonial silver carbonates, instead of antimonial silver sulphurets.

A prominent writer gives the following formula:
Silver Fahlerz, (argentiferous gray copper ore.)
Silver 17.71—31.29; antimony 26.63—24.63; Sulphur 23.52—21.17 Lustre, metallic; color, light steel gray.

In closed tube, sometimes decrepitates, melts and gives antimonial fumes and sulphurous acid.

On charcoal, it fuses easily and gives a bluish white coat of antimonious acid and antimonial fumes. There is also a yellowish coating close to the test, which appears white on cooling. This coating is created by oxide of zinc.

The Reese River ore, (Nevada,) from the Comet lode, seems to be a metamorphosed Silver Fahlerz. "The sulphur is represented by carbonic acid, so that almost all the copper and silver are classed as carbonates. It contains silver, 22.35; copper, 17; antimony and some lead. It has a dull greenish black or black color; streak shining; powder greenish gray. In a closed tube it yields nothing volatile. In an open tube some sulphurous acid can be observed. On charcoal fuses slowly, but boils up suddenly, in contact with glowing coal, leaving button of silver and copper. This button when played upon with the O. flame on another part of the charcoal, gives first a bluish coating of antimonious acid, then a yellow one, nearer to the assay of the oxide of lead. The silver can be separated from copper by cupellation with lead.

The silver fahlerz of the Sheba lode (Humboldt) contains : silver 8.20, gold o.oo8, some antimony and lead, but very little copper. It is called gray silver ore.

Each of these ores have the peculiarity similar to the Leadville and Silver Cliff ores.

CARBONATE OF LEAD.

A white salt forming beautiful rhombic crystals in nature, as cerussite, white lead, dutch ceruse, ceruse, flake white, krems white, etc.

It turns black by exposure to the action of sulphuretted vapors. Cerussite forms a vehicle for silver and the ores are known characteristically as soft carbonates. As a rule, silver is always combined with lead in lead deposites. Lead ores are found plentifully at Leadville, Colorado, both as galenas and carbonates.

The carbonate of lead is generally not rich in silver, but where it results from the decomposition of the argentiferous sulphurets, it usually contains more than the original ore. It has 68.3 per cent of lead. A transparent or translucent mineral, with a white or greyish color, that is sometimes tinged blue by the salts of copper ; crystallises in the trimetric system, but occurs also granular and massive, lustre, adamantine, inclining to vitreous ; streak. uncolored ; hardness 3 to 3.5, and specific gravity 6.46 to 6.48. Bp. decrepitates, becomes yellow, then red, and finally with care on charcoal, a globule of lead may be obtained. Dissolves in nitric acid with effervescence, and is also soluble in a solution of potash. This is the most abundant of the oxidised ores of lead. Frezenius says: "It is very slightly soluble in perfectly pure (boiled) water. 1 part requiring 50550 parts. It dissolves somewhat more readily in water containing ammonia and ammonium salts or carbonic acid. When ignited it loses its carbonic acid, leaving oxide. The ignited oxide slowly absorbs carbonic acid on exposure to the air. It turns moist litmus blue."

Its acid solution gives a fine yellow with chromate of potash or iodide of potassium.

This ore is often overlooked by the miner, or thrown aside as useless, its exterior resemblence to carbonate of lime, or earthy calamine rendering such deception very easy. The above chemical tests however will at once determine its value.

The lead being formed probably as a sulphate in the first place on the upheaval or the volcanic throes of the Earth's Interior and the intense combustion attendant thereon, the surface of the deposite has been subjected to the oxidising action of atmospheric air from which it extracted oxygen, producing what is known as the protoxide, having a dark surface covering; this again subjected to the moisture of external air acquired further oxygen together with carbonic acid, produced by the decomposition of wood, with its acetic acid forming the carbonate of lead. Place lead in contact with rotten wood, the pyroligneous and acetic acid will act upon it, forming a compound of the acetate and carbonate of lead.

Frezenius says :

" The sulphate of lead dissolves in **22,800** parts of cold water. It is readily soluble in hot potash or soda. Alkaline carbonate and bicarbonates (soda, potash, lime, or ammonia,) convert it, even at a common temperature, completely into carbonate of lead."

CARBONATE OF COPPER.

Mineral green, verditer green, mountain green, malachite. The blue carbonate is called azurite or blue verditer. .

Carbonate of soda added in excess to a solution of sulphate of copper, (blue vitriol) the precipitant is at first pale blue and flocculent, but a green tint and becomes the carbonate of copper.

The Blue carbonete has a Sp. gr. 3.5 to 3.7 hardness 3 to 4 Color from Azure to Berlin blue, with an occasional tinge of black. It yields easily to the knife. Before the Bp. it decrepitates, blackens aud ultimately fuses. It is soluble with effervescense in nitrie and the solution on the addition of ammonia in excess becomes azure blue. All solutions of copper, deposit that metal on a plate of clean iron.

The green carbonate has Sp. gr. 3 5 to 4, hardness 3.5 to 4, color various shades of green. Before Bp. it decrepitates and fuses into a black slag and behaves as the blue carbonates.

CARBONATE OF IRON.

Clay iron, spathic iron ore, sparry ore (crystallized) Sp. gr. 3.7 to 3 8, lustre, vitreous, streak or powder, white It has 38.63 acid ; 61.37 base.

This is protoxide of iron in combination with

carbonic acid and usually contains carbonate of manganese and magnesia. The color is white, yellowish, and often of a reddish hue or flesh colored, also brown varieties, partly oxides. This ore is usually mingled with pyrites and sulphurets of various descriptions.

The *argillaceous* variety, or compact carbonate of iron occurs chiefly in coal formations. It is found in round or flattened lumps, spheroids, imbedded in clay, clay slate, sand stone, and range from globules the size of peas to masses of two or more tons weight. It contains about 33 per cent of metal.

CARBONATE OF LIME.

Carbonate of lime takes a myriad forms such as marble, limestone, calcareous spar, rhomb spar, chalk, stalactites, stalagmites, oolite, Portland stone, shells, etc., etc. It has 43·71 parts carbonic acid, and 56.26 of base.

There seems to be a crust all over the Plains of Colorado, of soil impregnated with alkali varying in depth, from the edge of the mountains (where it is found at its greatest) to the eastern slope of the State where it is found in a much less degree. This is a powerful fertilizer and mixed with other soil or barnyard manure becomes one of the richest and most fertile soils, and is really the cause of the immense yield of wheat which has so astonished the agricultural world, and the plump fulness of the grains which is unequalled only where this description of fertilizer exists or is artificially provided. Its existence is owing to the washing of water in the mountains over the stratas of limestone which are found, sometimes in slaked and burned forms, taken up and carried away by the washings down on the Plains, on their passage extracting carbonic acid from the atmosphere, which is one of its ever present constituents, and by the time of its deposit becomes the carbonate of lime. In some places, the soil is so full of this deposit that it will not grow anything ; it burns up the plants ; this tendency can be neutralized by hauling manure or soil of a vegetable or alluvial character to mix with it, when it becomes of a most fertile character. These beds of alkali and carbonate of lime is really a redundancy of good, or too much of a good thing ; the presence of which in too great quantities is deleterious to vegetable life, for instance place a piece of pure pigeon ordure close to the root of a small plant of any kind and it will most probably kill it, take another piece of equal size and adulterate with soil and distribute it over a larger surface or mix it with water and use the solution as a liquid manure, over a larger surface and the value of good is very perceptible in an increased growth and quality.

CARBONATE OF SODA.

Soda, Barilla, etc.

This is more fusible than carbonate of potash, but not quite as soluble, therefore, makes a better flux and also a better glass.

Dry, it has 41.42 acid ; 58 58 base.

Crystallized, 15.43 acid ; 21.81 base ; water, 62.76.

Bicarbonate of soda, 58.58 acid ; 41·42 base.

CARBONATE OF POTASH.

Wood ashes, potash, grey salts, sub carbonate of potassa, salt of tartar, etc.

Ashes, 31.91 of acid and 68.06 base.

Bi-carbonate of potash has 48.38 acid; 51.62 base.

CARBONATE OF BARYTA.

A common natural product known as Witherite.

Sp. gr 4.3; hardness 3 to 3.5.

Exposed to Bp. flame in the platinum forceps, it melts readily into a white enamel, with a brilliant light. It dissolves in diluted hydrochloric nitric, and the solution gives a precipitate with solution of sulphate of lime or any other sulphate.

It is poisonous, has no effect on vegetable colors and is nearly insoluble in water.

Very plentiful at Leadville.

There are other carbonates, but which it would be a work of supererogation to speak of in this connection. It more properly belongs to the experienced mineralogist, and as this treatise is published more for the information of the miner and the masses, it would be best not to cumber our pages with such matter, which, on account of its technical character would not be read. References can be had to the various prominent writers on mineralogy for further information.

ORES OF TIN.

In connection with these carbonates, the ores of tin are nearly always found traversing the Elvans in Europe, and the same peculiarities may yet be found in Colorado. The only tin yet having been found occuring in the drift sands of the Arkansas and

Platte Rivers, and not very many miles from Lead-ville. We predict, therefore, for that reason and several other geological indications, that rich stan-niferous or tin ores will yet be found in or near the carbonate belt. as an oxide in cassiterite, and it would be well for the miner and prospector to be on the look out for the indications of its presence.

The small specimens of stream or wood tin which is found occasionally in the sand and gravel of the rivers, are but the rounded nodules of the ore as found in the veins. They are easily analysed by the assayer and reduced by the experienced metal-lurgist.

———

Extract from letter of A. A. Hodges of Nevada, Colo., October 1, 1865.

" For several years it has been the belief of miners that a belt of silver lodes passed through the coun-try, near the head waters of South Clear Creek, Snake, Blue, Platte and Arkansas Rivers. As far back as the year 1859, nuggets of silver were taken out of the Blue River and some of its tributaries; but not until within a year or two has any attention been paid as to the sources from whence these nug-gets came, but late investigations and assays have proven that we have in Colorado, some of the rich-est silver mines extant. I have not the least doubt but that the day is not far distant when the silver mines of Colorado, and especially those of Ten Mile District and other districts adjoining to it, will equal if not surpass any silver mines heretofore known.

I form this opinion from the results of a personal examination of that part of the country this summer, extendinrg through a period of several weeks.

The character of the ores and of the blossom rocks associated with the lodes, present unmistakable evidences of rich deposits of silver. I have heard of some assays as high as $7,000 per ton. The general course of the lodes is northeast and southwest, and in most instances crosses the hills in a direction well calculated for tunnelling. The width of the belt seems to be between 10 to 20 miles, and known as the Silver Belt of the Blue.

From some of the gulches in Ten Mile District a considerable quantity of pure silver nuggets have been taken averaging from an ounce to a pound in weight. These were picked up near the surface, and further developments must undoubtedly exhibit others of greater weight, The width of some of the crevices is enormous, in many instances ten and twenty feet, and even as high as forty feet. Ores are found upon the surface in abundance, requiring no delay, and but very little expense to supply abundant material for mills.

The lodes crossing the mountains assume the pitch of the stratified rock on the surface and become more vertical as they pass down into the eruptive granite, and in crossing the valleys become perpendicular.

———

Extract from letter of Judge Cowles, of Empire City, Colo., Oct. 9, 1865.

"As regards the Snake River Silver Mines, I think I have an acquaintance with them equal, perhaps, to that of any other person, having been one of the original discoverers of silver in that region, and having assisted in the first organization of the two mining districts there. I was there one year ago to-day, or rather on Ten Mile Creek, which particular district exhibits the perfection of silver lodes, in the rich out-croppings of galena.

You ask my opinion of the final results of mining in that country. My opinion is but that of a miner, and is, that I most positively believe that it is the richest silver mining district in the United States, if not in the world, and time will prove the truth of my assertion. My experience in silver mines has not been limited to Colorado; for I have been on the frontier for the last twenty years, and have visited the most profitable mining regions in the territories beyond."

Extract of report of A. A. Sawyer, dated at Central City, Colo., Oct. 10, 1865.

ORES.

Among the different kinds here visible, galena-argentiferous is most prominent, and in most of the mines thus far developed is the predominating element in the combined mass. The average assay of this ore has proven it to be very rich in silver, and some exceptional assays show a greater proportion of the precious metal than any one I have ever before known of.

The value, quantity and accessibility, comparatively speaking, of the silver element, have claimed for it primary attention, and the erection of properly constructed smelting furnaces cannot fail, in my opinion, to obtain results far beyond any yet obtained in this territory, so unprecedentedly rich in its mineral deposits.

Argentiferous galena is to the silver miner what gulch and placer diggings are to the gold miner; in the former case the galena from its fluxing property, furnishes an easy method for the concentration of the silver, while in the latter case, the thoroughly decomposed condition of the auriferous pyritous iron yields an equally sure method of abstraction. Some of the specimens of detached quartz upon being submitted to a strong heat, emit upon the surface many minute globules of pure silver, plainly visible to the naked eye. Many specimens of silver have been extracted from the ores of this region by the use of the blow pipe.

I consider the region rich beyond any expectations I had before visiting it.''

———

(*From Fuller's Treatise on Silver Mines.*)

''Wherever, in any part of the world, silver mines have been worked, they are worked now, unless arrested for some explainable cause.

The lack of machinery, the existence of war and the incursions of Indians, in Mexico, familiarized our minds with the idea of abandoned mines. But they have all been abandoned for some other cause than that they are exhausted. We know of no silver mining regions in the world which have given out.

The mines of Mexico, originally worked by the native Aztecs, before the Spanish conquest, are worked still. The mines of the Andes have given forth their wealth for more than three centuries. The mines of old Spain have been worked from the middle ages, and are in working condition now.

In Hungary. the same mines worked by the Romans, before the time of Christ, still yield their steady increase. The silver mines of Freiburg, in Saxony, worked from the eleventh century, have no diminution. In Bohemia, Tyrol, Norway and Sweden, in the Ural and Atlas Mountains, and, indeed, wherever the discoveries of silver have been made, we believe, without exception. the mines continue to be worked to the present day, and are generally more productive now than at any time during their past history, has had its parallel in no other businesss.''

ORIGIN

OF THE

FORMATION OF COAL.

Geology has established the fact, that from peat and forestal deposits our coal formations have sprung, but it has not hitherto enlisted the assistance of Chemistry to account for the various changes which the vegetable and woody material has undergone in its wonderful transition to form the fossil fuel of the Age of Man. Philosophers have agreed upon the generic and generalizing term of decomposition, but nothing farther. They have told about the possible generation of ulmic and carbonic acids and the possible carbonization by the latter.

Authorities agree that the formation of all the varieties of coal from the brownl ignite or Braunkohl of the Upper Tertiary to the anthracite of the Upper and Lower Carboniferous Periods are the results of the same process differing only in degree, but they say nothing of what that process is. It has ever been veiled in impenetrable mystery, and no one has attempted to solve the problem, save only with a vague indefinite generalization.

Dana says: " That, as yet, very little is known of its actual constituents."

All coal or fossil fuel is acknowledged to be of vegetable origin. Some years ago quite a Quixotic tilt was had between learned professors as to wheth-er coal was of vegetable or mineral origin. The intelligence of later discoveries and the advance-ment of Physical Science has placed the discussion beyond contradiction and resolved the proper posi-tion of the different coals. The distinguished mi-croscopist Professor Bailey, detected evidences un-mistakable, of the vegetable structure of the hardest authracite, and in fact, no one with any pretension to science, but will be prepared to add his testimony in confirmation of it.

Look at our peat depoits of to-day! In the Rocky Mountains near Empire City, and on the banks of the Platte River, on the Plains, are large peat bogs, overlying a clay substratum where the immense deposits of mountain flora such as ferns etc. have accumulated for centuries in our present epoch.

The enormous prolific vegetation of the carbo-niferous, Cretaceous and Tertiary Periods, furnishes ample fields for the study of the different classes of plants which form the coal deposits. The Bitumi-nous coals ot this region of Colorado, partaking as they do, of the conchoidal structure and fracture of the anthracite, mixed with the soluble resins and the product of the *Pinites Succifer* or Amber tree of the ancient forests which are found everywhere in their masses; fully and satisfactorily establishes their claims to a ligneous or woody structure common to the brown coal and the anthracites. The bitumi-nous coals of Pennsylvania are characterized by the same peculiarities as those of Colorado. We find our coals so erroneously classed as Lignites by some of the best professors and experts of the United States, lying side by side with the Oolitic stones of the Jurassic and the so called common black Lignite, the link between the brown coal and the bituminous, mixed with and associated with the Pyritous deposits of the Upper and Lower Cretaceous Strata.

Dr. Kane says :

" The wood has undergone a kind of decomposition."

He does not attempt to attribute the agency of anything but the internal heat of the Earth, to cause it.

Prof. Silliman says, in speaking of the decomposition of wood, when buried in the ground, and excluded from the action of the air :

" 'The oxygen which it contains gradually com-"bines with the carbon, to form carbonic acid, and "substances are obtained, in which the proportion "of carbon and hydrogen is greater than in the "original fibre. Peat, lignite and bituminous coal "are products of this decomposition."

Exactly, but the carbonic acid gas is evolved, which by sublimation forms the food of plants above, as is proved by the absence of any carbonates and the positive presence of sulphates in the coal. The learned Professor very shrewdly avoids saying, that coal is formed by the generation of carbonic acid.

The spontaneous combustion (as it is popularly termed,) of the peat and woody structure which forms the coal, is also demonstrated in the coal heaps in the vicinity of shafts ; here we find what we call spontaneous combustion in the interior of the slack pile, a stratum of it in some cases actually coked, caused no doubt by the decomposition of the bi-sulphuret of iron, that is, the change of the sulphuret into the sulphate (copperas) by the absorption of another equivalent of oxygen.

Strictly speaking. there is no such phenomenon as spontnaeous combustion. The inflammation of various organic and inorganic substances without the immediate contact of any ignited matter, which has given rise to the term, is, nevertheless, as certainly the result of some direct act or acts which can be accurately traced, as is the firing of a lucifer match when struck on a rough surface.

Dr. Buckland says;

" The flora of these strata (the leaves, peat, vegetable or woody tissues), underwent a course of chemical changes, and new combinations of their vegetable elements, converted them to the mineral condition, coal."

Not a word does this distinguished scientist tender to the world in common with all his scientific brethren of the chemical causes which lead to coal as a result.

Stockhardt says:

" They were buried under immense beds of clay and sand and were there decomposed by a process similar to that of putrefaction, while the sand hardened into sandstone and the clay into slaty clay or shale."

Not one word of any chemical change except from putrefaction. I dispute this theory of decay, there is no possible sign of decay or putrefaction ; but there is, at least to my view, from my standpoint, the most direct evidence of combustion, which is the antipodes of decay and putrefaction, though some chemists denominate decay as eramacausis or slow burning.

Dr. Fowne says :

"Some of the curious phenomena of communicated chemical activity, where a decomposing substance seems to involve others in destructive change, which without such influence would have remained in a permanent and quiescent state, and the actions attending thereto are yet very obscure and require to be discussed with great caution."

Sulphuric acid chars or blackens wood by abstracting a portion of its oxygen and hydrogen, the carbon being left in excess. This action of sulphuric acid is a consequence of its strong affinity for water, the elements of which it appropriates from most organic substances. Dextrine, sugar and many other substances can be manufactured from woody fibre by the agency of sulphuric acid. The tendency of wood to decay, is checked by the acids and certain salts, and promoted by the alkalies, which have the effect of hastening the decomposition of organic substances. Rotten stone is the production of calcareous material, operated upon by some acid, leaving the silica, alumina and carbon. This agent has been hitherto universally conceded to be the carbonic acid of the air, brought down by rain ; but Prof. Johnstone says in an able paper to the British Association in 1853, "but there were instances not capable of explanation by this agency alone, (speaking of carbonic acid) and attributable to other acids, which are produced under certain conditions, and exercise a much wider influence."

Prof. Johnstone seems to have had an indefinite indistinct idea of the true character of this agency, but failed to complete its accomplishment. The bottom of peat bogs and the substrata of the coal veins of Colorado present this same evidence, as spoken of in relation to the rotten stone, and that too very strongly of the action of acid ; the stone and clay are bleached and corroded, and only the silex and alumina left. Whole tree trunks exist imbedded in the gravelly detritus of streams near Denver, with emphatic evidences of corrosion and discoloration evidently formed by the combustion of acid. Thus the fire clay of the coal measures can be accounted for by the action of sulphurous acid developed in the vegetable deposit saturated with sulphur and iron, and thus forming the coal, the fire-clay sinking below.

Liebig, with that grand generalizing, but too bold dash at conclusions for which he is often guilty, says :

"The decay of vegetable fibre is a slow combustion or oxidation." I can not understand this statement, for decay is synonymous with putrefaction and so Stockhardt classes it, which is the very antipodes of combustion, known and admitted to be so, as well by Liebig, as by any other scientist. Decay may be oxidation, but not combustion.

Dr. Mc Farlane says :

" In anthracite coal, the process of liquefaction and carbonization or perhaps it should be called crystallization, has obliterated nearly all traces of the original vegetable matter."

Again he says :

" The vegetable matter must have been immediately covered with water as soon as it was formed, in order to be preserved from the rapid decomposition which always takes place in the open air."

Here again is a serious conflict of authorities.

Prof. Lesquereux says :

" We see it hardening (speaking of the woody matter in bogs), either by compression or by slow burning in water."

Dr. Mc Farlane, again says :

" The formation of different varieties of coal is supposed to be owing to the different degrees of progress made in the process of liquefaction and carbonization." Further, he says in conclusion : " The precise character of the process by which the change is brought about, may not be perfectly understood and there may here be room for farther examination to fully solve this interesting problem."

I can not believe the theory of decay or putrefaction, for where sulphuric acid is present, there can be no decay or putrefaction and we have abundant evidence of its presence in every kind of coal; but there must be with the agency of the heat generated from the interior of the Earth a very active and destructive decomposition or distillation of the woody tissues, and yet not by burning ; for there was evidently no burning, except the combustion of the woody fibre by the action of sulphurous media. There was no flame, only a charring, it is true, under a great pressure, as the coloration of the detritus above, below, and all around abundantly testifies.

The old theory of the slow process of the formation of coal, and the immense pressure, which has given some of our scientists such a magnificent opportunity for the exercise of their mathematical skill, has received a heavy blow from singular facts developed in the Shenandoah Valley and published in the *Shenandoah Herald*, in which is an account of the formation of anthracite coal from apparently pure spring water in a pipe used for draining the Indian ridge shaft of the Philadelphia and Reading Coal and Iron Company. It appears this coal forms in about four months by exposure to the air, thus scattering to the winds all the geological theories that coal takes thousands of years and heavy pressure to form it.

Coal is not sufficiently carbonized for commercial use, unless sulphur is used in its decomposition; but brown coal or lignite (the braunkohl of the German scientist) can be manufactured artificially in the laboratory without the use of any agency but heat and water; while black coal, the bituminous coal of Pennsylvania and Colorado can be made only by the aid of sulphur and iron. It is a well known, acknowledged and authenticated fact that in all black coals, sulphate of iron is present.

In our Colorado coals we find sulphate of lime streaks, adding another link in the chain of proof that sulphur was the decomposing element, added to a probable stream of water, which gave the oxygen for the formation of the sulphurous acid, moisten-ing the deposit, having passed over strata of lime-stone and iron.

We find the brown coal or lignite, where the structure of cedar wood in some places, the quercus and the palm in others lie side by side with the petrifactions of the same family having the same form, and in every way assimilating save in the car-bonization of the material.

In the neighborhood of Denver, Colorado, in the drift of a stream about 20 feet below the surface of the prairie I found whole trunks of charred or carbonized trees filled or saturated with iron pyrites and of the same character as the petrefactions in the same neighborhood snd evidently originating in the same local source.

Mr. G. C. Brodhead tells of similar instances, he says :

" Sticks of wood have often been found in modified drift, at 20 feet or more beneath the surface. In north Missouri sticks of wood have been found 75 feet from the surface, and part of a grape vine at 40 feet. In Illinois a piece of cedar at 100 feet. In Nevada, Mo., a walnut log, 2 feet thick, from 16 feet depth, and 4 miles north ; charred wood and bivalve shells at 19 feet deep."

These logs were evidently carried down from the summit of the Rocky Mountains by the torrents and floods which periodically pour down their Eastern slopes and form the gradually receding formation from the East front of the Sierras to the Gulf of Mexico.

I found also on the face of these drift banks on the shale and sandstone surrounding these logs, efflorescences of an impure alum, like what Dana says are found underneath a sheltered cliff of a stream in Nova Scotia, where it is evidently the result of the action of the decomposing pyrites. This efflorescence (specimens of which I have preserved) as also the wood pyrites (as it is called by curiosity dealers) has a very remarkable appearance, and I have never yet heard of any one drawing attention to it in this locality before, though large quantities can be found at any time.

I contend that the action of the iron and sulphur upon the vegetable matter of the wood generated sulphuric acid, forming under certain circumstances —potassa alum. soda alum, ammonia alum, chrome alum, or iron alum, forming not carbonates but sulphates in every case. This I think proves alone, to demonstration the theory I advance.

It is a well known fact that heaped-up iron pyrites in shale, when wetted, often causes the combustion of the pile, as in alum making, and has been used as an argument against the shipment of " brassy " coal, i. e.. coal containing these pyrites.

We find then, that leaves, peat, vegetable or woody tissues, saturated by sulphurous media as iron pyrites operated upon by means of heat, causing combustion and carbonization and thus accounts for the chemical formation of coal.

Sulphur takes fire at a temperature of 300° Fah., and forms sulphurous acid in combination with two volumes of oxygen to one of sulphur, or it forms sulphuric acid by one volume of sulphur vapor to six of oxygen. It does not form a direct union of oxygen and sulphur, but from the combination of sulphurous acid with another volume of oxygen.

It is a well known fact, demonstrated nearly half a century ago that sulphuric acid gas is invariably a product of the combustion of coal gas, forming another link in the chain of proof that sulphuric acid not carbonic acid gas was the cause of the formation of coal.

I have obtained a pure charcoal by the combustion of wood in sulphuric acid in the open air.

Dana says :

"Sulphur is present in nearly all coals. It is usually combined with iron."

Nearly every kind of coal is accompanied by a variety of pyrites and is known to experts as "coal brasses," and is often found in fragments and thin seams running parallel with the cleavage of the coal. It is a bi-sulphide of iron.

Dr. C. T. Jackson says :

"The manner in which the original potash has been removed from the vegetable substances, is not understood."

Adopt my theory of the formation of the coal measures and Dr. Jackson would not have made that remark, as he could then as I do now, easily account for and explain how the potash was removed from the original substances, viz : by its union with sulphur and the formation of sulphate of potassa and alumina.

Water does not assist in the combustion of coal, except where pyrites are concerned. There is much misunderstanding as to the part played by water in changes leading to combustion. The water itself is not decomposed, as some people have imagined. The heat evolved during the combustion of hydrogen and oxygen, during their combination to form water (the heat of the oxy-hydrogen blow-pipe) must be supplied before they can be again torn apart, so that, so far from water being a producer of heat, it is likely to be a consumer.

"Mix ¾ of an ounce of iron filings, ½ an ounce of flowers of sulphur, and ¼ of an ounce of water in a small vessel and put it in a warm place. The mass becomes heated, the water evaporates and in an hour a black powder is formed, in which no particle of iron or sulphur, is perceived. Sulphate of iron is formed."—*Stockhardt.*

After reading these various authorities and comparing or rather contrasting them with each other, indeed, we are led to the conclusion of Prof. Lesquereux when he says:

" The want of precise information on actual phenomena, the understanding of which is important for pursuit of geological studies, is perhaps nowhere more evident than in considering how little the formation of our combustible minerals is understood."

Water dissolves its equivalents of carbonic acid gas, and under pressure absorbs a very large quantity, a greater part of which is evolved under a relief from the pressure.

Carbonic acid will not under any circumstances either cause or support combustion.

Prof. James D. Dana, says in his exhaustive manual of Geology, when speaking of the decomposition of wood and leaves in the formation of coal.

" When the vegetable material is under water, the atmospheric oxygen is excluded, except the small part contained in water, and this oxygen with some proceeding from the growing plants in the waters, is all that comes from external sources. Under this diminished supply, part of the carbon and hydrogen escape oxidation and a coaly product is left behind. This covering of water prevents a complete combustion of the material, just like the covering of the earth over burning wood when charcoal is made. The air might also be partly or wholly excluded from vegetable debris, by a covering of clay or earth; and this is generally what happened, sooner or later in the Carboniferous Period.''

''The changes attending the ultimate decomposition under these circumstances depend on the affinity of (1) the carbon for oxygen, making carbonic acid; (2) of hydrogen for oxygen, producing water. ; (3) of carbon for hydrogen, making carbohydrogen gas or oil; and (4) on the tendency of the carbon and hydrogen under certain proportions, to form, with a portion of the oxygen, the stable compounds included under the term coal.'' The carbonic acid and water escape and also the carbohydrogen gas; and consequently under the most favorable circumstances, the wood loses in the change, much carbon and hydrogen as well as oxygen. It is probable in the making of bituminous coal at least 3·5 of the material of the wood are lost; and in the making of anthracite ¾. Besides this reduction to 2-5 and 1-4 by decomposition, there is a reduction in bulk by compression; which, if only to one half, would make the whole reduction of bulk to 1-5 and 1-8. On this estimate it would take five feet in depth of compact vegetable debris to make one foot of bituminous coal and eight feet to make one of anthracite. For a bed of pure anthracite 30 feet (like that at Wilkesbarre) the bed of vegetation should have been at least 240 feet thick.

This is certainly a most ingenious hypothesis of Prof. Dana, who is justly entitled to the position of the front rank of natural scientific savants. His equivalence of the gases is exceedingly fine, but he has not proven by any argument whatever that coal was produced by carbonic acid, he advances no modus operandi how the decomposition or the result could have been achieved, though he scientifically explains how certain gases could unite with others.

In regard to the reduction in bulk and thickness of the coal strata; his hypotheses are exceedingly shrewd, but they cannot be borne out by facts; I can take him to trunks of trees of various degrees of carbonization, retaining the rotundity of their external form, the diameter horizontally agreeing with the perpendicular, showing to demonstration that there was no reduction of bulk whatever. Again, in some coals the annular rings of the flora are distinctly traced, and in some instances the minutest delicate fronds of the ferns.

Prof. Lesquereux says:

"Chemistry accounts for the differences in the various degrees of decomposition of woody materials. It explains how the transformation of woody fibres into coal is the result of a retarded combustion by the slow combination of the oxygen of the atmosphere with the hydrogen of the plants, converting the woody fibre into carbon and increasing proportionally to the duration of the process the amount of fixed carbon."

I cannot agree with Prof. Lesquereux, though he echoes the theory of Liebig, that coal is the result of retarded combustion, but that it is a very active combustion,—so much so as to discolor and give character to the superincumbent strata, as also to that beneath it. It must be remembered that the heat of the atmosphere at that early period was tropical and the heat of the surface of the earth was much higher.

Again, I cannot understand how the slow combination of the oxygen of the atmosphere with the hydrogen of the plants can cause even the retarded combustion he speaks of; but if he add some iron pyrites to his combination, by that means generating sulphurous acid, I can then account for a very rapid combustion, as per my theory.

If the scientist examines the problem of the formation of coal through my theory, he will look at it from a very different, but at the same time a very elevated and comprehensive standpoint.

It is well known that the union of the sulphurous acid and iron, with the application of the heat, which at the period of the formation of coal was much greater than now, in presence of the oxygen within and enveloping the ligneous structure must have caused a generation of heat in the organic substance and the reagent, thereby producing combustion and altogether changing their character.

The theory of combustion has been discussed by the most eminent men of both this and the last century without bringing about any satisfactory explanation of the production of heat and light, and all chemists agree that it is but imperfectly understood. Lavoisier attempted, but failed most signally; Stahl with his phlogistic theory, supposed that a substance burnt, phlogiston escaped, leaving the substance incombustible; but in coal we see the fact which appears to be as patent as the sun at noonday, that oxygen (or dephlogisticated air, as Priestley called it) once upon a time supported the combustion, and heat was produced, but the product remained almost as combustible as it was in its initial structure.

Combustion ordinarily presumes the presence of

heat and light, but combustion can proceed without light, as in this case.

Lavoisier's theory of combustion, that oxidation was synonymous with it, in this place I do not wish to discuss, though it might possibly be borne out by my conclusions; but this theory of the formation of coal to my mind, settles many a knotty point which has appeared mysterious in the realms of chemistry in relation to this interesting subject.

And now let me state in conclusion, in my opinion, that the upheaval of the cordillera and the radical change in the position of the different rocks. caused the prostration and throwing off of the vegetable peat and forestal growth comprising vast areas of the luxuriant tropical growth of the carboniferous era, washed by the rains from off the scarps and counterscarps of the upheaved portions, collected and agglomerated in vast accumulated deposits in the canons and low places at the foothills and bases of the mountains. These accumulations in time became saturated with water, impregnated with iron, forming the oxide of iron, running in its course over the lava beds and beds of sulphur thrown up from the depths of the earth, forming the sulphurets of iron, which by combustion under pressure of the continuous deposit of sand above the vegetable accumulation, and the addition of a tropical heat below converted the sulphurets of iron into sulphurous acid; collecting other equivalents of oxygen, it formed sulphuric acid.

ASSAYING

BY

SCORIFICATION & CUPELLATION.

ASSAYING OF METALLIC ORES.

" The term assaying (Fr. *Essayer*, to try,) is applied always to metallic compounds. "

Before metallic ores are worked upon in the large way, it is necessary to ascertain what kind of metal and how much of it is contained in a proportionate quantity of the ore, to learn whether it can be economically worked and the acknowledged best systems of reduction practised upon it, to extract the precious or valuable metals, and in what manner the process can be conducted to accomplish that end. The knowledge requisite for this is called the docimastic art.

SAMPLING AND PULVERIZING.

A judicious selection in the first place should be made and an assay made of each one from the valuable vein matter, and one from the gaugue immediately on each side of it. This will give you a pretty good idea of the economic value of the ore deposite and the mine.

Great care should be taken that a fair average sample of the ore is selected as that is the base and foundation of the whole. The quantity, of course, must be governed by circumstances, but it is always best to get as much as a pound, so that a complete and thorough admixture and average can be obtain- ed. This is the "secret of success." If the sam- ple weigh, only one ounce, as much care must be taken in the sampling as if it weighed ten pounds. One portion may be very rich or very poor, it must always therefore be a fair average. If the substance is wet, it should be dried in an earthen saucer with gentle heat in the muffle, taking care not to roast or calcine it. Very hard specimens should be brok- en up on the anvil and if refractory and inclined to scatter fragments, should be wrapped in paper or cloth to prevent loss of part of the ore, which might be the richest. This is especially applicable to rocks containing free metals or metallic gold or silver alloys. In some cases a cold chisel must be used to cut metallic specimens for assay.

Place the ore in an iron mortar, first cleaning it by rubbing clean sand in it with the pestle, carefully washing, and wiping with a cloth or brush. Strike the larger pieces of the sample by direct blows with the pestle, and when the whole is reduced to a size equal to small sand, transfer it to a porcelain mor- tar and submit it to the process of rubbing, not striking as in the case of the iron mortar, until it is finely pulverized and capable of passing through a sieve of 60 to 80 meshes to the linear inch. In general, the more finely is the substance pulverized, the more accurate and expeditious the assay. It is always best to cover the mortar with a damp cloth, not only to prevent loss of pieces of the ore, but also for the purpose of confining the fine dust, which in some cases is highly injurious if inhaled.

Some minerals can be pulverized with greater ease if they are heated and suddenly quenched with cold water: such as charcoal, mica, flint, and the silicious gangues,—as gold quartz.

If particles of metallic silver or malleable ore remain upon the sieve, they must be assayed separately.

An extempore sieve can be made of a piece of fine lawn or muslin; tie it up loosely and shake the powder upon a piece of paper or pane of glass, where it is submitted to the process of sampling or quartering, which is done by placing the powder on a sheet of cardboard, pane of glass or other plane level surface, and with a knife divide it into four parts, any three of which parts can be discarded and the fourth part taken and again divided into four other parts, which should be repeated again if the quantity of the sample will justify it.

RECORDING THE ASSAY, ETC.

The next step is placing the pulverized and quartered sample in a box, or other receptacle, properly labelled and numbered. so that it can be identified at any future time, if necessity require. It is usual in well-regulated offices for specimens of each uncrushed sample to be kept labelled and filed away, especially those which run high in the noble metals. The numbers should also agree with a schedule of assays, kept as a permanent record. with all the information necessary for future reference. Such a record should embrace date, number of assay, name of owner, district, name of lode, letter of box, what metals to reduce, amount charged for assaying, result in weight, result in U. S. gold, and general remarks.

Some assayers keep a number of round boxes, well made of good material, lettered from A to Z, and at the end of every day or every week's work, transfer the powder to a neat paper envelope, and file away with the number properly marked as in the schedule or record book.

PREPARING THE FIRE.

This is in itself an art, requiring just as much care, method and system as any other part of the assay, or much time may be lost.

Plenty of paper or shavings should first be placed in the furnace, upon which good small pine sticks are placed together with a quick-kindling hard coal or charcoal on the top, and when the coal gets red, feed with coke if coke is to be chiefly used. Coke to light a fire is not good, as it requires a blast to start it, which is not ordinarily appended to a muffle furnace. Take care to break up the coal and coke into small enough pieces to readily fall down between the muffle and the furnace sides. Care should also be taken to keep the flue at the muffle as free as possible, to enable the oxidizing fumes to pass off readily.

Light the fire.

Shut the mouth of the muffle.

Close the grate door.

Keep open the ash-pit door until the fire is well lighted and the coals are red, then close it to heat the muffle.

Get the muffle to as high a heat as possible. You cannot get it too hot for scorification. It can only be raised to a bright red heat.

WEIGHING THE SAMPLE.

Assayers vary in the amount of the ore they take for assaying. Some take ⅓, others ⅙, or 1-10th or 1-20th of the assay ton.

I prefer 1-10th of an assay ton for a charge.

The assay ton equals 29.166 grammes, or 29⅙— or in round numbers, 450 grains.

1-10 of an assay ton therefore equals 45 grains
1-6 " " " 75 "
1-3 " " " 150 "

If the ⅓ or ⅙ of an assay ton be used, and the ordinary charge of an assay ton of granulated lead, it makes the lead button too large for the ordinary cupel, and if 12 parts of lead are used as in copper, arsenical or antimonial ores, the button is much too large, not reliable, and apt to lose the precious metal by running over, or being absorbed with the litharge into the body of the cupel. 1-10th of an assay ton therefore is much the most convenient charge to use.

OBJECT OF SCORIFICATION.

The object is to concentrate all the silver or gold in a lead button, to desulphurize the ore, to place the gangue and earthy matter with the oxide of lead into a slag, which, when cool, can be knocked off and separated.

The chemistry of this process is, that the sulphur of the heavy metallic sulphides passes off as sulphurous acid, while sulphides of the alkalies and alkaline earths, if present, are oxidized to sulphates.

The great object of scorification is to oxidize the metals alloyed with the gold and silver, by means of the protoxide of lead (litharge), which has the peculiar property of fusing, and forming fusible compounds with nearly all the metals except gold, silver, platinum and mercury.

PREPARING THE SCORIFIERS.

Get good scorifiers, number them with the Roman Numerals I, II, III, IV, V, VI, VII, VIII, IX, X, etc., and note the number in the record. See that the scorifiers are good or the assay will go through them and be lost.

Place the sample (1-10 of an A.T.) previously carefully weighed, in the scorifier and weigh out for ordinary ores one A. T. of granulated lead. The following table is considered the standard for the quantity of lead required.

Galena requires 6 parts lead and no borax; quartzose ores about 8 parts and no borax, blende, mispickel and pyrites about 16 parts, and $\frac{1}{4}$ to 1 part borax; copper and tin 20 to 30 of lead; nickel and cobalt even more; nickelspeisse 16 of lead and repeated scorifications, ores containing calcite, dolomite, barytes or fluorspar 8 of lead and 12 parts borax or glass. In cases of doubt as to the nature of the ore commence with 10 parts of lead and if the fusion is not good, repeat with a larger proportion of lead.

Put $\frac{1}{8}$ of the lead in the bottom of the scorifier to prevent the ore sticking; intimately mix the charge of ore with $\frac{3}{8}$ more of the lead and place it on the other lead in the scorifier, add the remaining 4-8 or half of the lead on the top as a cover, and upon the top, place borax glass sufficient in powder to cover a dime or lump about the size of a pea.

If borax glass cannot be had, use common borax, some common flint glass should be mixed with it.

The granulated lead used, should be known to be chemically pure; if it is not, it is necessary to make an assay of it separate from the others, and subtract the amount of silver from the product of the other assays.

THE PROCESS.

Get the muffle to a pretty high heat and keep it up for the purpose of melting the lead.

Put the scorifier into the hottest place in the muffle.

Keep the mouth closed, and regulate the draft carefully excluding all currents of air.

The borax glass will melt into a globule.

Then the lead will fuse and envelope the mass. It is first covered by a play of beautiful iridescent colors resulting from the formation of a thin pellicle of the oxide of lead. This is soon supplanted by a yellow scum of the same substance; this scum should be broken by a hot wire, and placed to the side of the scorifier in order to expose a fresh sur-face for oxidation, or place ½ inch square block of pine wood.

When heated to redness the metal gives off distinct vapors which burn in the air with a white vivid flame, carried away through the orifice at the back of the muffle.

Fuse until the ring of slag closes completely over the lead, then raise the heat as high as you can to liquefy the slag and settle the lead.

54

About fifteen or twenty minutes is generally sufficient to accomplish the melting or fusion of the lead, if you start with a good hot muffle. With refractory ores it may take thirty minutes for complete fusion, and even then it may be necessary to add some more borax or lead.

If the slag is thick and does not easily flow, it requires the addition of more lead in small quantities.

If necessary to add more lead it should be wrapped in a piece of paper and dropped on the melted mass. The paper keeps it from contact with the assay until its water is driven off, thus preventing loss by spitting.

Sulphur gives light gray; zinc, thick white; arsenic, grayish; and antimony, bluish flames.

If the ore contains much zinc, it can be volatilized by covering the scorifier with glowing coals, closing the muffle and increasing the heat, as oxide of zinc forms a stiff slag.

It is always best to carry the scorification as far as possible as experience has shown that there is less loss of silver in scorification than in cupellation.

POURING INTO THE MOULD.

When the scorification is completed, then take the scorifiers, one by one, commencing at the first and each number successively and pour the contents quickly into the iron moulds, which generally are cast together to the number of twelve or less or more, taking the precaution of chalking the cavities, so that the metallic buttons will not adhere to the iron.

If no mould or casting plate is at hand, the buttons may be left in the scorifiers to cool. When cool, the metallic buttons should separate easily from the slag. It can be freed by hammering on an anvil into a cube .

If the button is too large it must be reduced in bulk by re-roasting in a fresh scorifier.

If it is too hard and not easily malleable or contains too much copper, add more lead and borax and repeat the scorification.

CUPELLATION.

By this means the slags are absorbed by the porous mass of bone dust forming the cupel and thus expose the clean surface of the hot metal to the oxygen of the air. All the metals which can be oxidized under the influence of oxygen and heat are thus oxidized and absorbed by the cupel. Those which cannot be oxidized and remain after the application of the strongest heat on the cupel are called precious metals in contradistinction to the others, which are called base metals.

CUPELS.—These are ordinarily made of white bone ashes. A variety of materials can be used, but none superior. A good cupel is a most important auxiliary in the dry assay. It is essential that the black matter in burning bones should be expelled. The ashes should be perfectly white without a shade of gray, when moistened. The burned bones may be crushed in a mortar or ground in a mill. In all cases they must be converted into a fine powder, passed through a fine silk sieve and washed in lukewarm water, which removes the soluble salts in the ashes. It may be repeated to free the powder entirely from such salts, as they absorb moisture, retain it, and crack the cupels when exposed to the heat. The remaining powder consists chiefly of phosphate of lime, mixed with a little carbonate of lime and some silicious matter derived from the ashes of charcoal.

The cupels are made in a circular mould made of steel, iron or brass, and can be obtained in any city when required, of the requisite size, taking care that they are large enough for a good sized lead button, the small size sometimes made is too small. Sometimes larger cupels are required, of two inches in diameter, these should be surrounded by an iron ring and put altogether into the muffle, one ring will serve for some fifteen or twenty charges. Without the ring the cupel is apt to crack and the assay lost.

The moistening of the ashes in making cupels requires great care and caution. If the ashes are too damp the cupel will be porous and liable to absorb metal with the oxides. If too dry, the cupel will be too dry, not porous enough and will work slow, requiring a high heat to absorb the oxides. The best time for making cupels is when the ashes have been washed with warm water and are thoroughly wet. In this damp state, they are exposed in a warm room, or to the rays of the Sun and constantly stirred to prevent particles becoming too dry. The mass is dried in this manner until it will hardly adhere together when squeezed in the hand. It is then wrapped in paper and surrounded by a damp cloth to prevent evaporation. Some manufacturers use beer or dissolved starch for glueing the ashes together. To give the cupels when made, a certain degree of firmness, a little carbonate of potassa is sometimes added to the water for moistening the bone-ash. The amount usually required is exceedingly small, The size of a large nut is sufficient to a pint operation. If beer or starch is used the alkali is not used. When this way of damping the ashes is adopted, the cupels must be exposed to a red heat before they are fit for cupellation. Such admixtures cause the cupels to be porous, and for alloys of gold. silver and particularly copper, too porous. Good cupels consist of ashes neither too fine or too coarse; fine ashes are required for alloys of gold and silver. The cupels for pure gold or pure silver may be made of coarser ashes than those for alloys. In assays of minerals, we hardly ever know if the refined metal is pure or alloyed : it is therefore necessary to use the finest kind of bone ashes and use more time in cupelling. Coarse ashes cause the cupels to be weak and liable to break ; and what is worse, such cupels absorb metal as well as oxides. All the disadvantages resulting from the use of fine ashes are slow work, more fuel and more time ; but the assay is always more correct than in cupels made of coarse ashes.

When lead only is to be removed, the cupel of bone dust is the best.

When copper, iron, arsenic and similar metals are in the alloy, this cupel does not give at first a correct assay, and it is necessary to add more lead and cupel again. In such instances it has been found that cupels made of two parts of wood ashes and one of lime-stone marl are preferable to bone ashes. In the mode of manufacturing the cupel there is no difference ; the ashes worked the same, and the same process throughout, with the addition, that a thin layer of bone ashes is placed over the cavity in the mould before the ramrod is driven. To prevent cracks in the cupel, the concave surface should be turned downwards until used.

Cupelling is one of the most interesting operations in metallurgy. The chief condition of success is, that the newly-formed oxides, of which those of lead and bismuth are the only perfect ones, should be absorbed by the cupel. Copper is also, but not in a large quantity. It is mechanically conducted to the pores, and soon fills the surface of the cavity, after which no more is absorbed. In a limited quantity all the oxides are absorbed by the cupel, but not directly by the mass of the cupel. These oxides are conducted to the pores by the oxides of lead or bismuth. As the latter metal is very scarce and cannot be generally obtained, we confine our remarks to lead only.

The cupellation is performed generally on a number of tests at once, for it causes as much labor to refine one test as a muffle full of them.

Those specimens which contain gold, platinum, copper, iron and other substances, are placed further in the muffle when the strongest heat prevails. The tests which are alloys of pure silver and lead, or antimony, may be cupelled near the mouth. If we neglect to attend to this, it may happen that one is frozen while the heat carries off from the other the silver or gold or both together, by evaporation. A frozen cupel must be removed to a hotter place.

The cupels are marked with red chalk, with a number corresponding with the record book and the balance of the ore not used and filed away.

When the furnace is well under weigh, insert the cupels, and when they have attained a white heat, the lead button or test is gently laid in the cavity, but not before it is held for a short time over the cupel to heat it. Some assayers put the test in the cupel before they put it into the muffle, but it should not be, for a cupel may look well when cold, but on heating it, the invisible cracks open, and cause a loss of the assay by absorbing both metal and slag.

Before putting metal in the hot cupel, it should be closely examined by bringing the eye to the mouth of the muffle, where it is protected against the radiating heat by a pane of glass or mica. If no cracks are visible, the metal may be placed in the cupel. A cupel will absorb twice its weight of litharge or oxide of lead, but it is not advisable to put more than its own weight into it, for an excess of lead will filter through the cupel and break the bottom of the muffle. To meet this difficulty, it is always advisable to keep a good layer of ashes on the floor of the muffle to absorb any lead which may be forced out of or through the cupel.

The cavity in the cupel should not be too deep, for this causes ashes above the test ; besides, the oxide of lead does not easily rise above its level and a deep cupel works slowly as the fresh air which enters the muffle and passes over the cupel does not enter it. In most cases, we have an alloy for cupellation in the form of a solid button of lead, which is placed with tongs in the proper cupel. If the test is in fine grains or in dust, such as a precipitate of gold it is wrapped in a piece of thin sheet lead, and put into the cupel. In this latter case it is advisable to melt first a little pure lead in the cupel before putting in the test.

In case a test is frozen, the assay would be lost, if we did not furnish the frozen cupel with a fresh supply of pure hot lead, which will gather the metal and afford an opportunity of recovering the test.

When all the cupels are occupied, fill the mouth of the muffle with a piece of tough or knotted well burned charcoal. The lead soon melts and if the heat is sufficiently strong, a cloud of white vapors of lead rises over the cupels. If the cloud is low— that is—hovers over the surface of the cupels—indicates that the muffle is too cold. The mouth should be closely shut, the fire stirred to increase the heat. By a small hole, between the coals at the mouth of the muffle we may now observe its interior. When the cloud of lead smoke rises about half-way between the roof and the bottom of the muffle, the heat is strong enough; we now remove a small coal, and admit more air to the interior. When the heat increases so as to cause the vapors of lead to ascend to the roof, the coals at the mouth should be removed, by this means the cloud is lowered. If the heat is too high, the hot oxide of lead, filters through the cupel and destroys the muffle; it also evaporates silver and gold whose vapor passes off with those of the lead.

If the heat is too low, the oxides formed do not melt at once and penetrate the cupel; a part of the precious metals is carried over into the pores of the vessel. When the operation is well conducted the melted test shows a clear white color. with a metallic lustre, and on its surface we observe a constant circular motion, unaccompanied by any sign of ebullition. When nearly all the lead is evaporated and the test reduced to a size which contains about two parts of lead to one of silver, the mouth of the muffle is closed once more, and the heat increased. After a few minutes, we observe the globule becomes quiet and assumes a clear mirror like surface. no motion is perceptible, and around it there is a rose of dark oxides in case the alloy contains any copper, iron or other refractory metals. If the heat is strong and no vapors any longer rise, the globule has become clear ; we then gradually remove the coal from the mouth of the muffle and draw the cupels gently towards it by means of a hook of strong iron wire. If we remove the hot cupels directly from the muffle into the cold air, the hot globules are liable to explode, particularly when silver predominates in the alloy. The surface of the globule under these circumstances, cools rapidly and chills, while the gases which may be within breaks the crust in endeavoring to escape and throw off the parts already cooled and solid. When the globules are very small, or contain much gold, they will explode and the cupels, may be removed at once with the tongs.

If the test lead used is not quite pure ; if it contains silver; which most lead does, we take an amount of it, equal to that used in the assay and cupel it by the side of its corresponding test. This is called a witness, or lead test. The silver thus obtained from this witness is subtracted from the assay and the difference is the actual yield of the assay.

It is best to commence cupelling under a strong heat and modify it in the course of the process. If the test is poor in silver, the heat may be stronger throughout the operation than if it is rich. The last heat should be strong in all cases, but of short duration, to prevent as much as possible the evaporation of silver.

If it is found necessary to cool either cupel, it may be done by placing a piece of cold iron or clay beside it. If the whole muffle is too hot, and the removal of the coal at its mouth does not reduce the heat sufficiently, a shovel or a piece of cold iron held within will check it

The color of the cupel shows the kind of metals in the test besides lead and precious metal.

Pure lead colors a cupel of white bone ashes, yellow, which is often inclined to orange.

Bismuth does the same.

Copper causes a gray coating, which is often reddish or brown.

Iron produces a black, often brownish coating.

Tin causes a gray one and invariably freezing.

Zinc deposits a yellow powder and causes freezing; and occasions also a loss in silver, by evaporation and ebullition.

Antimony causes a bright yellow color, and in most cases a considerable loss in silver by evaporation: It also causes the cupel to crack, which then absorbs the metal.

When the cupellation is almost finished, a play of colors is seen, and it suddenly assumes a brilliant color, which is called "the brightening of the assay." When the silver is freed from all the lead, then move it to the mouth of the muffle so as to cool slowly. If it is suddenly cooled, it will, what is called "sprout" or vegetate, and loss of silver may ensue: This is a peculiarity of pure silver, caused by the sudden escape of the oxygen contained within it.

WEIGHING THE BUTTON.

Upon the completion of the cupellation, the round, bright, smooth button bead lying in the cupel must be carefully detached by a fine pair of pincers, pliers or forceps, and slightly pressed so as to free it from superfluous litharge or oxide of lead. It is advisable, sometimes, to flatten the bead on a small steel anvil to clean it. Examine with a glass to be satisfied that it is perfectly clean and ready for weighing.

Then place it in the pan of the balance, after testing the balance to see if it is perfect in adjustment. In the other pan put the necessary weights, carefully selected with a pair of fine forceps, never using the fingers Draw down the glass slide in front, to exclude all dust, wind, etc ; gently raise the beam by using the small clamp screw in front of the case ; note the oscillation of the perpendicular needle over the dial plate, together with the number of parts it passes over. By that means a fair approximation can be made as to its relative weight, even as low as the 10th of a millegramme. See that the pointer covers the exact centre of the dial Never attempt to estimate by appearance or judge by the oscillation from side to side, but always make an unalterable rule to keep adding or taking out weights until the pointer permanently settles to its place of proper equipoize. Then, if your scales are accurate, your weight is absolutely correct

In weighing ores and buttons. never buy cheap weights; always obtain the best. It has been noticed even in new weights, sometimes, that in a pennyweight there has been a difference of two grains from the true standard.

The balance also is most important, as there is a big joke told of an assayer of a prominent company not a hundred miles from Denver, who found in the tailings of the mill. more silver than he found in the original ore. He said, it was the fault of the balance (?) Therefore it is always best to provide good tools to work with, then there is no excuse for charlatans.

PARTING.

The separation of the metals: It is done by the humid or wet process, by nitric acid and aquaregia.

After the weighing, enough silver should be added to the combined silver and gold button, sufficient to make the proportion in value of three silver to one of gold. Then add about ten times or more the weight of dilute nitric acid, boil gently for a short time, say fifteen or twenty minutes, sufficient to dissolve the silver. Then decant and wash ; add strong nitric acid to dissolve the last portion of silver, which the dilute acid failed to dissolve. Boil a few minutes more, then wash dry and heat to redness in a dish ; the dark precipitate is gold, which must undergo a second weighing, thus obtaining the amount of the gold ; which, subtracted from the first weighing, gives also the weight of the silver.

APPENDIX.

Table of Troy Weights used with Gold and Silver, and Platina.

24 grains (gr.) make 1 pennyweight (dwt.)
20 pennyweights — 1 ounce (oz.)
12 ounces — 1 pound (lb.)

lb.	oz.	dwt.	gr.
1 =	12 =	240 =	5760
	1 =	20 =	480
		1 =	24

The value of gold is given in carats fine, 24 c. f. being pure.

One pound of gold 24 carats fine contains 5760 grs. of pure gold, as gold of that number of carats fine is unalloyed.

One pound of gold 23 carats fine contains 5520 grs., and so on, and one oz. of gold 20 c. f. has 400 grs.

one dwt. of gold 15 c. f. has 15 grs.

These are given as examples of the manner in which the amount of the pure metal ought to be calculated.

Table for Reducing Troy American to French Metric Weights.

GRAINS.	MILLIGRAMMES.	GRAINS.	GRAMMES.
1-75 equal to	.8632	20 equal to	1.2958
1-70 "	.9256	30 "	1.9437
1-65 "	.9968	40 "	2.5916
1-60 "	1.0799	50 "	3.2395
1-55 "	1.1780	60 "	3.8875
1-50 "	1.2958	70 "	4.5354
1-45 "	1.4398	80 "	5.1833
1-40 "	1.6198	90 "	5.8313
1-35 "	1.8512	100 "	6.4791
1-30 "	2.1597	200 "	12.9583
1-25 "	2.5916	300 "	19.4375
1-20 "	3.2395	400 "	25.9167
1-15 "	4.3194	500 "	32.3959
1-10 "	6.4791	600 "	38.8751
1-5 "	12.9583	700 "	45.3543
1-4 "	16.1979	800 "	51.8335
1-2 "	32.3959	900 "	58.3137
1 "	64.7919	1000 "	64.7919
2 "	129.5838	2000 "	129.5838
3 "	194.3757	3000 "	194.3757
4 "	259.1676	4000 "	259.1676
5 "	323.9595	5000 "	323.9595
6 "	388.7514	6000 "	388.7514
7 "	453.5433	7000 "	453.5433
8 "	518.3352	8000 "	518.3352
9 "	583.1371	9000 "	583.1371
10 "	647.9192	10,000 "	647.9192

French Weights reduced to English.

		Troy Grains.	lb.	oz.	dr.	gr.
Milligramme,	(1,000th of a grm. =	.0154				
Centigramme,	(100th of a grm. =	.1543				
Decigramme,	(10th of a grm. =	1.5434				
Gramme,	(Unit of weight) =	15.4340				
Decagramme,	(10 grms.) =	154.342 =	0	0	2	34.34
Hectogramme,	(100 grms.) =	1,543.4023 =	0	3	1	43.4
Kilogramme,	(1,000 grms.) =	15,434.0234 =	2	8	1	14.02
Myriagramme,	(10,000 grms.) =	154,340.2344 =	26	9	4	20.23

Grm.	Troy Grains.	Decagrm.	Dr.	Grains.	Hectogrm.	Troy Ozs.
1 =	15.434	1 =	2	34.34	1 =	3.2165
2 =	30.868	2 =	5	8.68	2 =	6.4330
3 =	46.302	3 =	7	43.02	3 =	9.6493
4 =	61.736	4 =	10	17.36	4 =	12.8660
5 =	77.170	5 =	12	51.7	5 =	16.0825
6 =	92.604	6 =	15	26.04	6 =	19.2990
7 =	108.038	7 =	18	0.38	7 =	22.5165
8 =	123.472	8 =	20	34.72	8 =	25.7330
9 =	138.906	9 =	23	8.06	8 =	28.9495

Various Modes of Distinguishing Minerals.

The miner and prospector should by all means accustom himself to examine minerals and ores by their physical properties before he proceeds to test their value by assay or analysis.

The text books and manuals lay down so plainly the different scales, numbering each according to its degree or ratio of standard, that with a very little practice any one can soon become expert in the discrimination.

Hoping that if this treatise will touch the note of vibration, the miner will follow up and dive deeper into the subject, we give a few of the physical properties of minerals and their standard of degree.

Physical Properties of Minerals.

A mineral is any substance in nature not organized by vitality, or that does not grow.

The popular belief has it, that stones and minerals grow, but they do not in the same mode as wood grows ; hence the distinction. They increase in size, but only by the addition of particles to their external structure.

Minerals have various characteristic peculiarities.

They have different colors.

Their hardness is in different degrees or ratios from being soft, or easily indented by the nail, to the hardness of the diamond.

Weight, from extreme lightness to extreme heaviness.

Lustre, from a total absence of the power of reflecting light, to the brilliancy of the mirror.

Transparent as glass in some, opaque as others.

A few have taste.

In their structure, they have cleavage, look at mica, for instance; the feldspars have the same peculiarity plainly developed. Granite and iron have a cleavage but not so plainly marked as the feldspars and the micas.

Crystallization is another peculiarity of minerals, which has been reduced to a science of itself, each class of minerals having distinct forms of crystals and are known by their angles and their varied forms.

Minerals are also distinguished from each other by their subjection to the action of heat and to acids and reagents.

One mineral melts with heat, others do not.

Some are infusible, others evaporate. Chemical analysis pursues this mode of investigation to a very great extent.

Scale of Hardness.

The comparative hardness of minerals should be carefully examined by the student. As standards of comparison, the following have been acknowledged universally:

1. Talc, soapstone or tailor's chalk; easily scratched by the nail.

    Lead is 1.5; mercury is 1; sulphur 1.5—2.5.

2. Rock salt, gypsum, (crystalized) zinc, copperas; not easily scratched by the nail, and does not scratch a copper coin.

Silver is 1.5 — 3 ; gold is the same ; copper 2.5—3 ; bismuth is 2— 3.5 ; spathic iron, baryta 2.5 — 3.5.

3. Calc spar, transparent variety (carbonate of lime) ; scratches and is scratched by a copper coin.

Antimony is 3—3.5.

4. Fluorspar, crystalized variety ; not scratched by a copper coin, does not scratch glass.

Platinum is 4—4.5 ; tin is 4—5 ; iron 4—5, iron pyrites 5—5.5.

5. Apatite transparent crystal ; scratches glass with difficulty, easily scratched with the knife.

Nickel 5—6 ; cobalt 5—6 ; limonite, iron ore 5—5.5.

6. Adularia (feldspar) cleavable variety ; scratches glass easily, not easily scratched by the knife.

7. Quartz (rock crystal) transparent variety ; not scratched by the knife, yields with difficulty to the knife, scratches glass with facility.

8. Topaz, transparent variety ; harder than flint.

9. Emerald, corundum or ruby, sapphire, cleavable variety, harder still. Manganese 6—10.

10. Diamond, highest in the scale ; will scratch or cut any other substance.

Lustre of Minerals.

1. Metallic or the lustre of metals. Imperfect metallic lustre is known as sub-metallic. Arsenic is sub-metallic ; sulphur has a metalloid lustre.

2. Vitreous, or the lustre of broken glass. Imperfect vitreous lustre is known as sub-vitreous.
   Quartz is vitreous.
   Calcareous opal is sub-vitreous.
3. Resinous, like opal or broken resin.
4. Pearly, like pearl.
5. Silky, like silk, asbestos, fibrous gypsum, etc.
6. The lustre of the diamond.

Degrees of Fusibility.

One of the most important modes of discrimination of minerals, after the other modes just mentioned have been examined, is the degree of their fusibility. Von Kobell has arranged a table of minerals in their order of fusibility, which may assist the student in the modus operandi.

1. Stibnite (gray sulphuret of antimony.)
2. Natrolite.
3. Almandine (garnet, carbuncle.)
4. Strahlstein (actinolite.)
5. Adularia (feldspar.)
6. Diallage (bronzite.)

If a mineral be more fusible than No. 2, and less than No. 3, it is said to be 2.5 in the scale, etc.

Multiplication Table for Gold.

$20.67 \times 1 = 20.67$	$20.67 \times 4 = 82.68$	$20.67 \times 7 = 144.69$
$20.67 \times 2 = 41.34$	$20.67 \times 5 = 103.35$	$20.67 \times 8 = 165.36$
$20.67 \times 3 = 62.01$	$20.67 \times 6 = 124.02$	$20.67 \times 9 = 186.03$

Table of Melting Degree Points of Various Metals.

	Centigrade.	Fahrenheit.
Mercury	39 5	39
Sulphur	109	208
Potassium	62	144
Sodium	97.8	230
Tin	228	442
Cadnium	228	442
Bismuth	264	507
Lead	325	627
Zinc	412	773
Antimony	621	1150
Bronze	900	1652
Silver	1000	1832
Copper	1091	1996
Gold	1100	2012
Steel	1300	2372
Bar Iron	1500	2732
Cast Iron	1530	2786

Fusibility of Metallic Compounds.

Parts Zinc.	Parts Lead.	Parts Bismuth.	Melts at Deg. Fah.
2	3	5	199
1	1	2	200.7
3	5	8	202
3	3	8	202
3	6	3	208
3	8	8	226
2	4	4	243
6	1	5	245
1	1	1	254
4	5	4	256
4	6	4	270
1	0	1	286
45.3	0	54.6	325
3	2	0	333
2	1	0	360
8	0	1	392
1	3	0	500

Reduction of the Carbonates.

In these ores there is found combined sulphur, arsenic and antimony, (and which it is essentially necessary should be volatilized.) The great fault of metallurgists is a failure to reduce ores economically. If they are rich ores the large margin may leave room for grave mistakes in reduction ; but in low grade ores or such that run below twenty-five or thirty dollars, there is no margin for mistakes and practical metallurgists and those too who understand their business, alone, should attempt to handle them. Ores can be worked at a price not to exceed fifteen dollars per ton if care is taken.

The sulphur can easily be eliminated, and almost as easily also can the arsenic, but the antimony being more refractory, is no doubt the shoal or sunken reef upon which so many metallurgists fail in Colorado.

Metallurgists, commonly, are not chemists, and they seem not to understand the principles involved and the importance attached to a knowledge of the fusibility of metals, and more particularly of the carbonates.

The carbonates of Leadville are especially adapted to chlorination as also to amalgamation with chloritic roasting, requiring a rather prolonged roasting to thoroughly throw off the carbonic acid gas and form a soluble chloride for either leaching or amalgamation.

The galena ores are best adapted to the smelting process by the cupola furnace.

Formulas.

BLACK FLUX.—Potash forms an important adjunct in the tests, fluxes and reagents, to assist in the reduction of refractory metals and earthy minerals. They are all combinations of this salt. The black flux is made by mixing 2 parts of powdered tartar (bitartrate of potash,) and one of nitrate of potash. This mixture is deflagrated in small portions at a time in an iron ladle or crucible, when it becomes of a black color.

WHITE FLUX.—This is made by mixing one part nitre and two parts cream of tartar, or crude argol.

CORNISH REFINING FLUX.—Deflagrate and afterwards pulverize two parts nitre and one part tartar.

CORNISH REDUCING FLUX.—Mix well together, ten ounces tartar, three ounces and six drams nitre, and three ounces and one dram borax.

To Refine Cement Copper.

This process has for its object the purification of the oxide of copper or a coarse copper button, the result of a previous roasting. It rests on the property possessed by copper when in a melted condition, and exposed to the action of heated air, of remaining unoxidized as long as iron, lead, arsenic, bismuth, antimony, tin, zinc, nickel and cobalt are present. These metals are turned into oxides in such a case, and when borax is at hand, are taken up and carried away as a fusible slag.

The button is placed in a previously heated crucible or scorifier with an equal weight of borax, in the muffle, surrounding it with glowing coals to a high red heat. The button melts in a few minutes and by opening the door of the muffle a stream of air is allowed to pass over the fluid. The oxides of the metals form, collect on the surface and produce a play of colors; and at length the button having lost much of its impurity, becomes greenish, usually sinks beneath the borax. and on account of the higher smelting point of copper. becomes solid. The cup is now taken from the muffle and cooled by dipping it slowly into water. The copper is found in a button among the slag; it should be well formed, bright and malleable, and when possessed of these qualities may be at once weighed.

Another Process.

100 parts of metal, 10 parts of copper scales, and 10 parts ground bottle glass or other similar flux. After the copper has been kept in fusion for half an hour, it will be found at the bottom of the crucible perfectly pure, while the iron, lead, arsenic &c., with which it was combined, will be oxidized by the scales and dissolved in the flux or volatilized. A pure copper has thus been obtained from brass, bell metal, gun metal and several alloys containing from 4 to 50 per cent. of iron, lead, antimony, bismuth, arsenic &c.

To Separate Silver from Copper.

Granulate the copper, dissolve in sulphuric acid, make green vitriol by crystallization, or precipitate with iron. Silver is the residue in the vats undissolved.

Fluxes.

Limestone, feldspar, fluorspar, calcite, lead, coal dust, charcoal dust, quartz, sand, salt, slate, volatile alkalies, pitch, iron filings, slags &c. are used according to the nature of the ores.

Iron ores, on account of the argillaceous or clayey earths they contain, require calcareous (chalk or lime,) additions.

Copper ores require slags, vitrescent or glassy additions, like borax or common glass.

The proper mixing of ores and fluxing them is one of the most important things in metallurgy, and therein lies the secret of the smelter's success, as in the mechanical manipulation ot alloys lies also the refiner's success.

Fluxes for Metallic Sulphurets.

Antimony.—Iron, lime and black flux, by calcination.

Lead.—By roasting.

Iron.—By lime.

Cobalt and Nickel Arseniurets.—By roasting, arsenious acid expelled, oxide of cobalt or nickel produced and remaining.

Copper.—By calcination and arrest of process before the iron fusion point is reached. The separation of the iron is effected by adding sand before reduction ; the silicic acid unites with oxide of iron. Copper is reduced.

Silver.—By Pattinson process and cupellation.

Sulphurets are generally fluxed in a small way by fusion, with a mixture of lime and charcoal, or of

carbonate of potash and charcoal or black flux.—
Kane.

Copper and iron pyritous ores with but little silica,
are difficult to reduce alone, and require quartz or
quartzoze rock as a flux.

An ore containing sulphur 45, iron 40. copper 6
to 10, balance silica, water, etc , requires 20 per
cent. of silica or quartz in order to combine with
the iron, after its liberation from the sulphur to
reduce it to regulus.

If the metallurgist wishes to obtain formulas of
any kind for the manipulation of ores, they can be
furnished by the author on application. Our limited
space will not allow of a complete discussion of such
an extensive subject.

Wet and Dry Parting of Gold and Silver.

These processes are difficult to operate except to
experienced experts, therefore we will omit them,
giving our space to more necessary formulas, trust-
ing to find probably in some future publication,
more space than we can allow here. The wet parting will be explained under the *modus operandi* of
assaying.

To Detect Gold in Minerals.

Scrape the mass with the point of a knife; if it
be gold, it will be soft and may be cut like lead; or
strike it gently with the small end of a hammer, if it
be gold, it will be indented. Melt a small piece
with blowpipe, if gold, its color remains; but if it
be brittle and hard to the knife and hammer, it is

not gold. Place a few fragments upon a hot shovel or in the flame of a blowpipe, if the sulphur burn away, leaving scoria that is attracted by the magnet, this proves that it is a combination of sulphur and iron, or iron pyrites. Put a few particles into a watch glass and drop a little muriatic acid upon it, and hold it over the flame of a candle until it boils, if gold, no alteration; but if not, effervescence and change of color will result. If no change take place with hydrochloric acid, add *aqua Regia.* A solution is now sure to ensue, to which the usual tests for gold may be applied.

To Detect Silver.

A rich ore will be soft to the knife or hammer, and melt under the blowpipe with little difficulty, and by repeated fusion with borax, a bead of silver may be produced. A few particles of ore may be put into a watch glass, into which, drop a little nitric acid; then hold it over the flame until it is dissolved. After this dilute it with water, and stir it about with a bright copper wire; if any silver is present it will precipitate upon the copper, covering it with silver. Or add a little table salt to the solution, a white cloud of chloride of silver will fall down.

To Detect Copper.

Place piece of ore upon charcoal with a little borax and fuse with blowpipe. If rich ore it will be reduced to a bead of pure copper, coloring the slag green or reddish-brown ; it is sometimes necessary to repeat the fusion.

Another Method.—Reduce to powder, put in a watch glass with a few drops of nitric acid, and if no action takes place apply a little heat. The acid will dissolve the copper. Add a few drops of water and stir with point of a knife or any piece of clean iron. The copper will be deposited on the iron.

To Detect Lead.

Break a small portion from the ore and observe the fragments and their brilliancy ; now place a piece not larger than a pepper corn on charcoal and with a blowpipe blow a flame from a lighted candle, directing the jet upon the ore. If it contain lead it will discharge sulphurous vapors and in half a minute, the lead will be reduced. The ores of this metal are numerous ; the most common is blue lead ore, which occurs in great quantity. Others are of various colors, as gray, green, brown, yellow and red.

Blowpipe—Characteristic Tests.

Potassa, colors the flame violet ; best seen through a blue glass.

Soda, reddish yellow flame.

Lithia, carmine red flame.

Ammonia, colors red litmus-paper blue, pungent odor.

Baryta, burnt with alcohol gives a yellowish-green flame.

Strontia, crimson flame.

Lime, colors the flame feebly red, becomes caustic and glows when heated.

Magnesia, gives with nitrate of cobalt a pale flesh-color, after long blowing.

Alumina, gives a fine blue color, with nitrate of cobalt.

Silica, in S. Ph. bead, gives a semi-transparent skeleton floating in the bead.

Oxide of antimony, on charcoal, is reduced, and gives white fumes and coat and greenish-blue flame.

Arsenious acid, with soda on charcoal, gives white fumes and garlic odor.

Oxide of bismuth, on charcoal, is reduced to metal, and gives an orange-yellow color.

Oxide of cadmium, coats the coal with a reddish-brown powder and variegated tarnish.

Oxide of chromium, with soda in the O. F. gives a yellow glass; in R. F , green on cooling.

Oxide of cobalt, on charcoal becomes magnetic. With borax and S. Ph. beads, Smalt blue glass.

Oxide of copper, metallic button on charcoal. With borax bead, green glass, blue when cold; red in R. F.

Oxide of gold, with borax on coal, easily reducible to metal.

Oxide of iron, on coal becomes magnetic. Borax bead, red to yellow on cooling; in R. F., bottle-green.

Oxide of lead, reducible to metal on charcoal with sulphur-yellow coat and blue flame.

Oxide of manganese, with soda, on cooling bluish-green. With borax, amethyst bead, colorless in reducing flame.

Oxide of mercury, volatile on charcoal, metallic mirror with soda in closed tube.

Molybdic acid, with S. Ph., yellowish-green, and colorless when cold. The bead on coal becomes green on cooling.

Oxide of nickel, on charcoal, yields a magnetic powder. Borax bead, reddish-brown.

Oxide of silver, on charcoal, reducible to metal. With borax, opalescent or milk-white glass.

Oxide of tin, reducible on charcoal to metal with yellow coat, white when cold. With cobalt solution in O. F., gives a bluish-green color.

Titanic acid, with salt of phosphorous bead in R. F., a fine violet color.

Oxide of zinc, yellow coat on coal, white when cold. With cobalt solution. green in O. F.

Chlorine, with oxide of copper in borax bead, a fine azure blue flame.

Iodine, with soda, or better, bi-sulphate of potash; in matrass, violet fumes.

Bromine, with bi-sulphate of potash in matrass, reddish-yellow vapors.

Fluorine, etches glass when mixed with a little sulphuric acid.

Carbonic acid, acid reaction, turns lime-water white.

Sulphur, burns on charcoal with a blue flame with odor of sulphurous acid; better in open tube.

Ductility and Malleability of Metals.

Metals. Ductile and Malleable.	Brittle Metals.	Metals in Order of Wire Drawing Ductility.	Metals in Order of Laminable Ductility.
Cadmium	Antimony	Gold	Gold
Copper	Arsenic	Silver	Silver
Gold	Bismuch	Platinum	Copper
Iron	Cerium	Iron	Tin
Iridium	Chromium	Copper	Platinum
Lead	Cobalt	Zinc	Lead
Magnesium	Columbian	Tin	Zinc
Mercury	Iridium	Lead	Iron
Nickel	Manganese	Nickel	Nickel
Osmium	Molybdenum	Palladium	Palladium
Palladium	Osmium	Cadnium	Cadnium
Platinum	Rhodium		
Potassium	Tellurium		
Silver	Titanium		
Sodium	Tungsten		
Tin	Uranium		
Zinc			

Silver Ores and Minerals.

Minerals.	Contents.	Per. Ct. of Silver.
1. Naumannite	Silver, Selenium	73.2
2. Eucairite	Silver, Selenium, Copper	43.1
3. Hessite	Silver, Tellurium	62.8
4. Sylvanite	Silver, Tellurium, Gold	10-15
5. Silver Glance	Silver, Sulphur	87.04
6. Stromeyerite	Silver, Sulphur, Copper	2 96-13.1
7. Sternbergite	Silver, Sulphur, Iron	34 2
8. Miargyrite	Silver, Sulphur, Antimony	36 7
9. Pyrargyrite	Silver, Sulphur, Antimony	59 8
10. Proustite	Silver, Sulphur, Arsenic	65 4
11. Stephanite	Silver, Sulphur, Antimony	68.5
12. Brogniardite	ilver, Sulphur, Antimony, Lead	26 1
13. Polycasite	Silver, ulphur, Antimony, Copper, Arsenic	68
14. Tetrahedrite	Silver, Sulphur, Antimony, Copper, Iron, Zinc, Mercury, Arsenic, Bismuth	3.09-31.29
15. Xanthoconite	Silver, Sulphur, Arsenic	64.
16. Fireblende	Silver, Sulphur, Antimony	62.2
17. Freislebenite	Silver, Sulphur, Antimony, Lead	22-24
18. Kerargyrite	Silver, Chlorine	75.33
19. Bromegrite	Silver, Bromine	57.4
20. Embolite	Silver, Bromine, Chlorine	61-71
21. Iodorite	Silver, Iodine	46.

Table of Elements and Combining Weights.
PROF. CHANDLER.

		OLD.	NEW.				OLD.	NEW.
Aluminum	Al	13.7	27 4	Mercury	Hg		100.	200.
Antimony	Sb	122.	122.	Molybdenum,	Mo		48.	96.
Arsenic	As	75.	75.	Nickel	Ni		29 4	58.8
Barium	Ba	68.5	137.	Nitrogen	N		14.	14.
Bismuth	Bi	210.	210.	Osmium	Os		99.6	199.2
Boron	B	11.	11.	Oxygen	O		8.	16.
Bromine	Br	80.	80.	Palladium	Pd		53 3	106.6
Cadmium	Cd	56.	112.	Phosphorus	P		31.	31.
Cæsium	Cs	133.	133.	Platinum	Pt		98.7	98.7
Calcium	Ca	20.	40.	Potassium	K		39.1	39·1
Carbon	C	6.	12.	Rhodium	Ro		52.2	104.4
Cerium	Ce	46.	92.	Rubidium	Rb		85.4	85.4
Chlorine	Cl	35.5	35.5	Ruthenium	Ru		52.2	104.4
Chromium	Cr	26.1	52.2	Selenium	Se		39.7	79 4
Cobalt	Co	29.4	58.8	Silicon	Si		14.	28.
Columbian	Cb	94.	94.	Silver	Ag		108.	108.
Copper	Cu	31.7	63.4	Sodium	Na		23.	23.
Didymium	D	47.5	95.	Strontium	Sr		43.8	87.6
Erbium	E	56.3	112.6	Sulphur	S		16.	32.
Fluorine	F	19.	19.	Tantalum	Ta		182.	182.
Glucinum	Gl	46.	92.	Tellurium	Te		64.	128.
Gold	Au	197.	197.	Thallium	Ti		204.	204.
Hydrogen	H	1.	1.	Thorium	Th		115.7	231.5
Indium	In	56.7	113.4	Tin	Sn		59.	118.
Iodine	I	127	127.	Titanium	Ti		25.	50.
Iridium	Ir	99.	198.	Tungsten	W		92.	184.
Iron	Fe	28.	56.	Uranium	U		60.	120.
Lanthanum	La	46.	92.	Vanadium	V		51.3	51.3
Lead	Pd	103.5	207.	Yttrium	Y		30.8	61.7
Lithium	Li	7.	7.	Zinc	Zn		32.6	65.2
Magnesium	Mg	12.	24.	Zirconium	Zr		44.8	89.6
Manganese	Mn	27.5	55.					

Metals, Characteristics.—Ricketts.

Metal.	Color.	Hardness.	Best Solvents.	On Charcoal before the Blp.	Fusibility Fah.
Lead	Bluish, malleable	1.5	Nitric or Muriatic	Fuses and gives a yellow coat	234
Antimony	Bluish-white, brittle	3—3.4	Aquaregia	Fuses and gives off white fumes	425
Silver	White, malleable	2.5—3	Nitric and sulphuric	Fuses gives reddish coat with long blowing	916—1040
Gold	Yellow, malleable	2.5—3	Aquaregia	Fuses to a button	2016
Platin'm	Whitish to steel-gray, malleable	4—4.5	Aquaregia	Infusible	In flame of oxy h Blp
Zinc	Bluish-white, malleable brittle	2	All acids	Oxydizes and gives a white coat	412
Mercury	Tin-white, liquid	1—	Nitric	Valatilizes	Solid at 40 5
Bismuth	Reddish to silver- w'it brittle	2—3.5	Nitric	Fuses and gives an orange yellow coat	268.3
Tin	Like silver more bluish, malleable	4—5	Muriatic Sulphuric	Gives metallic Globule and white coat	235
Copper	Red, malleable	2.5—3	Conc acids	Can be fused to a bead	1090
Iron	Gray malleable, magnetic	4—5	All acids	Infusible	Highest heat of forge 1500—1600
Manganese	Grayish-white, brittle	0—10	Nitric sulphuric muriatic	Infusible	"

Metal, Characteristics—Continued.

Metal	Color.	Hardness.	Best Solvents.	On Charcoal before the Blp.	Fusibility Fah.
Nickel	Silver-w'it malleable, magnetic	5—6	Nitric	Infusible	Highest heat of forge 1500--1600
Cobalt	Steel-gray to red, magnetic	5—6	Nitric	Infusible	"
Carbon	Colorless to black	Variable	Insoluble	Infusible—burns	Infusible
Sulphur	Yellow, reddish, greenish, brittle	1—2.5	Oil of Turpentine, etc.	Melts and gives off sulphurous acid	111—114

Order of Conduction in the Principal Metals.

1. Silver,
2. Gold,
3. Copper,
4. Aluminium,
5. Zinc,
6. Iron,
7. Tin,
8. Platinum,
9. Lead.

Order of Electrical Conducting Power.

1. Silver = 1000
2. Copper = 999
3. Gold = 777
4. Zinc = 290
5. Platinum = 180
6. Iron = 168
7. Tin = 123
8. Lead = 83
9. Bismuth = 12

GOLD RUSH BOOKS

OREGON, USA

www.GoldMiningBooks.com

Books On Mining

Visit: www.goldminingbooks.com to order your copies or ask your favorite book seller to offer them.

Mining Books by Kerby Jackson

Gold Dust: Stories From Oregon's Mining Years - Oregon mining historian and prospector, Kerby Jackson, brings you a treasure trove of seventeen stories on Southern Oregon's rich history of gold prospecting, the prospectors and their discoveries, and the breathtaking areas they settled in and made homes. 5" X 8", 98 ppgs. Retail Price: $11.99

The Golden Trail: More Stories From Oregon's Mining Years - In his follow-up to "Gold Dust: Stories of Oregon's Mining Years", this time around, Jackson brings us twelve tales from Oregon's Gold Rush, including the story about the first gold strike on Canyon Creek in Grant County, about the old timers who found gold by the pail full at the Victor Mine near Galice, how Iradel Bray discovered a rich ledge of gold on the Coquille River during the height of the Rogue River War, a tale of two elderly miners on the hunt for a lost mine in the Cascade Mountains, details about the discovery of the famous Armstrong Nugget and others. 5" X 8", 70 ppgs. Retail Price: $10.99

Oregon Mining Books

Geology and Mineral Resources of Josephine County, Oregon - Unavailable since the 1970's, this important publication was originally compiled by the Oregon Department of Geology and Mineral Industries and includes important details on the economic geology and mineral resources of this important mining area in South Western Oregon. Included are notes on the history, geology and development of important mines, as well as insights into the mining of gold, copper, nickel, limestone, chromium and other minerals found in large quantities in Josephine County, Oregon. 8.5" X 11", 54 ppgs. Retail Price: $9.99

Mines and Prospects of the Mount Reuben Mining District - Unavailable since 1947, this important publication was originally compiled by geologist Elton Youngberg of the Oregon Department of Geology and Mineral Industries and includes detailed descriptions, histories and the geology of the Mount Reuben Mining District in Josephine County, Oregon. Included are notes on the history, geology, development and assay statistics, as well as underground maps of all the major mines and prospects in the vicinity of this much neglected mining district. 8.5" X 11", 48 ppgs. Retail Price: $9.99

The Granite Mining District - Notes on the history, geology and development of important mines in the well known Granite Mining District which is located in Grant County, Oregon. Some of the mines discussed include the Ajax, Blue Ribbon, Buffalo, Continental, Cougar-Independence, Magnolia, New York, Standard and the Tillicum. Also included are many rare maps pertaining to the mines in the area. 8.5" X 11", 48 ppgs. Retail Price: $9.99

Ore Deposits of the Takilma and Waldo Mining Districts of Josephine County, Oregon - The Waldo and Takilma mining districts are most notable for the fact that the earliest large scale mining of placer gold and copper in Oregon took place in these two areas. Included are details about some of the earliest large gold mines in the state such as the Llano de Oro, High Gravel, Cameron, Platerica, Deep Gravel and others, as well as copper mines such as the famous Queen of Bronze mine, the Waldo, Lily and Cowboy mines. This volume also includes six maps and 20 original illustrations. 8.5" X 11", 74 ppgs. Retail Price: $9.99

Metal Mines of Douglas, Coos and Curry Counties, Oregon - Oregon mining historian Kerby Jackson introduces us to a classic work on Oregon's mining history in this important re-issue of Bulletin 14C Volume 1, otherwise known as the Douglas, Coos & Curry Counties, Oregon Metal Mines Handbook. Unavailable since 1940, this important publication was originally compiled by the Oregon Department of Geology and Mineral Industries includes detailed descriptions, histories and the geology of over 250 metallic mineral mines and prospects in this rugged area of South West Oregon. 8.5" X 11", 158 ppgs. Retail Price: $19.99

Metal Mines of Jackson County, Oregon - Unavailable since 1943, this important publication was originally compiled by the Oregon Department of Geology and Mineral Industries includes detailed descriptions, histories and the geology of over 450 metallic mineral mines and prospects in Jackson County, Oregon. Included are such famous gold mining areas as Gold Hill, Jacksonville, Sterling and the Upper Applegate. **8.5" X 11", 220 ppgs. Retail Price: $24.99**

Metal Mines of Josephine County, Oregon - Oregon mining historian Kerby Jackson introduces us to a classic work on Oregon's mining history in this important re-issue of Bulletin 14C, otherwise known as the Josephine County, Oregon Metal Mines Handbook. Unavailable since 1952, this important publication was originally compiled by the Oregon Department of Geology and Mineral Industries includes detailed descriptions, histories and the geology of over 500 metallic mineral mines and prospects in Josephine County, Oregon. **8.5" X 11", 250 ppgs. Retail Price: $24.99**

Metal Mines of North East Oregon - Oregon mining historian Kerby Jackson introduces us to a classic work on Oregon's mining history in this important re-issue of Bulletin 14A and 14B, otherwise known as the North East Oregon Metal Mines Handbook. Unavailable since 1941, this important publication was originally compiled by the Oregon Department of Geology and Mineral Industries and includes detailed descriptions, histories and the geology of over 750 metallic mineral mines and prospects in North Eastern Oregon. **8.5" X 11", 310 ppgs. Retail Price: $29.99**

Metal Mines of North West Oregon - Oregon mining historian Kerby Jackson introduces us to a classic work on Oregon's mining history in this important re-issue of Bulletin 14D, otherwise known as the North West Oregon Metal Mines Handbook. Unavailable since 1951, this important publication was originally compiled by the Oregon Department of Geology and Mineral Industries and includes detailed descriptions, histories and the geology of over 250 metallic mineral mines and prospects in North Western Oregon. **8.5" X 11", 182 ppgs. Retail Price: $19.99**

Mines and Prospects of Oregon - Mining historian Kerby Jackson introduces us to a classic mining work by the Oregon Bureau of Mines in this important re-issue of The Handbook of Mines and Prospects of Oregon. Unavailable since 1916, this publication includes important insights into hundreds of gold, silver, copper, coal, limestone and other mines that operated in the State of Oregon around the turn of the 19th Century. Included are not only geological details on early mines throughout Oregon, but also insights into their history, production, locations and in some cases, also included are rare maps of their underground workings. **8.5" X 11", 314 ppgs. Retail Price: $24.99**

Lode Gold of the Klamath Mountains of Northern California and South West Oregon
(See California Mining Books)

Mineral Resources of South West Oregon - Unavailable since 1914, this publication includes important insights into dozens of mines that once operated in South West Oregon, including the famous gold fields of Josephine and Jackson Counties, as well as the Coal Mines of Coos County. Included are not only geological details on early mines throughout South West Oregon, but also insights into their history, production and locations. **8.5" X 11", 154 ppgs. Retail Price: $11.99**

Chromite Mining in The Klamath Mountains of California and Oregon
(See California Mining Books)

Southern Oregon Mineral Wealth - Unavailable since 1904, this rare publication provides a unique snapshot into the mines that were operating in the area at the time. Included are not only geological details on early mines throughout South West Oregon, but also insights into their history, production and locations. Some of the mining areas include Grave Creek, Greenback, Wolf Creek, Jump Off Joe Creek, Granite Hill, Galice, Mount Reuben, Gold Hill, Galls Creek, Kane Creek, Sardine Creek, Birdseye Creek, Evans Creek, Foots Creek, Jacksonville, Ashland, the Applegate River, Waldo, Kerby and the Illinois River, Althouse and Sucker Creek, as well as insights into local copper mining and other topics. **8.5" X 11", 64 ppgs. Retail Price: $8.99**

Geology and Ore Deposits of the Takilma and Waldo Mining Districts - Unavailable since the 1933, this publication was originally compiled by the United States Geological Survey and includes details on gold and copper mining in the Takilma and Waldo Districts of Josephine County, Oregon. The Waldo and Takilma mining districts are most notable for the fact that the earliest large scale mining of placer gold and copper in Oregon took place in these two areas. Included in this report are details about some of the earliest large gold mines in the state such as the Llano de Oro, High Gravel, Cameron, Platerica, Deep Gravel and others, as well as copper mines such as the famous Queen of Bronze mine, the Waldo, Lily and Cowboy mines. In addition to geological examinations, insights are also provided into the production, day to day operations and early histories of these mines, as well as calculations of known mineral reserves in the area. This volume also includes six maps and 20 original illustrations. **8.5" X 11", 74 ppgs. Retail Price: $9.99**

Gold Mines of Oregon - Oregon mining historian Kerby Jackson introduces us to a classic work on Oregon's mining history in this important re-issue of Bulletin 61, otherwise known as "Gold and Silver In Oregon". Unavailable since 1968, this important publication was originally compiled by geologists Howard C. Brooks and Len Ramp of the Oregon Department of Geology and Mineral Industries and includes detailed descriptions, histories and the geology of over 450 gold mines Oregon. Included are notes on the history, geology and gold production statistics of all the major mining areas in Oregon including the Klamath Mountains, the Blue Mountains and the North Cascades. While gold is where you find it, as every miner knows, the path to success is to prospect for gold where it was previously found. **8.5″ X 11″, 344 ppgs. Retail Price: $24.99**

Mines and Mineral Resources of Curry County Oregon - Originally published in 1916, this important publication on Oregon Mining has not been available for nearly a century. Included are rare insights into the history, production and locations of dozens of gold mines in Curry County, Oregon, as well as detailed information on important Oregon mining districts in that area such as those at Agness, Bald Face Creek, Mule Creek, Boulder Creek, China Diggings, Collier Creek, Elk River, Gold Beach, Rock Creek, Sixes River and elsewhere. Particular attention is especially paid to the famous beach gold deposits of this portion of the Oregon Coast. **8.5″ X 11″, 140 ppgs. Retail Price: $11.99**

Chromite Mining in South West Oregon - Originally published in 1961, this important publication on Oregon Mining has not been available for nearly a century. Included are rare insights into the history, production and locations of nearly 300 chromite mines in South Western Oregon. **8.5″ X 11″, 184 ppgs. Retail Price: $14.99**

Mineral Resources of Douglas County Oregon - Originally published in 1972, this important publication on Oregon Mining has not been available for nearly forty years. Included are rare insights into the geology, history, production and locations of numerous gold mines and other mining properties in Douglas County, Oregon. **8.5″ X 11″, 124 ppgs. Retail Price: $11.99**

Mineral Resources of Coos County Oregon - Originally published in 1972, this important publication on Oregon Mining has not been available for nearly forty years. Included are rare insights into the geology, history, production and locations of numerous gold mines and other mining properties in Coos County, Oregon. **8.5″ X 11″, 100 ppgs. Retail Price: $11.99**

Mineral Resources of Lane County Oregon - Originally published in 1938, this important publication on Oregon Mining has not been available for nearly seventy five years. Included are extremely rare insights into the geology and mines of Lane County, Oregon, in particular in the Bohemia, Blue River, Oakridge, Black Butte and Winberry Mining Districts. **8.5″ X 11″, 82 ppgs. Retail Price: $9.99**

Mineral Resources of the Upper Chetco River of Oregon: Including the Kalmiopsis Wilderness - Originally published in 1975, this important publication on Oregon Mining has not been available for nearly forty years. Withdrawn under the 1872 Mining Act since 1984, real insight into the minerals resources and mines of the Upper Chetco River has long been unavailable due to the remoteness of the area. Despite this, the decades of battle between property owners and environmental extremists over the last private mining inholding in the area has continued to pique the interest of those interested in mining and other forms of natural resource use. Gold mining began in the area in the 1850's and has a rich history in this geographic area, even if the facts surrounding it are little known. Included are twenty two rare photographs, as well as insights into the Becca and Morning Mine, the Emmly Mine (also known as Emily Camp), the Frazier Mine, the Golden Dream or Higgins Mine, Hustis Mine, Peck Mine and others. **8.5″ X 11″, 64 ppgs. Retail Price: $8.99**

Gold Dredging in Oregon - Originally published in 1939, this important publication on Oregon Mining has not been available for nearly seventy five years. Included are extremely rare insights into the history and day to day operations of the dragline and bucketline gold dredges that once worked the placer gold fields of South West and North East Oregon in decades gone by. Also included are details into the areas that were worked by gold dredges in Josephine, Jackson, Baker and Grant counties, as well as the economic factors that impacted this mining method. This volume also offers a unique look into the values of river bottom land in relation to both farming and mining, in how farm lands were mined, re-soiled and reclamated after the dredges worked them. Featured are hard to find maps of the gold dredge fields, as well as rare photographs from a bygone era. **8.5″ X 11″, 86 ppgs. Retail Price: $8.99**

Quick Silver Mining in Oregon - Originally published in 1963, this important publication on Oregon Mining has not been available for over fifty years. This publication includes details into the history and production of Elemental Mercury or Quicksilver in the State of Oregon. **8.5″ X 11″, 238 ppgs. Retail Price: $15.99**

Mines of the Greenhorn Mining District of Grant County Oregon - Originally published in 1948, this important publication on Oregon Mining has not been available for over sixty five years. In this publication are rare insights into the mines of the famous Greenhorn Mining District of Grant County, Oregon, especially the famous Morning Mine. Also included are details on the Tempest, Tiger, Bi-Metallic, Windsor, Psyche, Big Johnny, Snow Creek, Banzette and Paramount Mines, as well as prospects in the vicinities in the famous mining areas of Mormon Basin, Vinegar Basin and Desolation Creek. Included are hard to find mine maps and dozens of rare photographs from the bygone era of Grant County's rich mining history. **8.5″ X 11″, 72 ppgs. Retail Price: $9.99**

Geology of the Wallowa Mountains of Oregon: Part I (Volume 1) - Originally published in 1938, this important publication on Oregon Mining has not been available for nearly seventy five years. Included are details on the geology of this unique portion of North Eastern Oregon. This is the first part of a two book series on the area. Accompanying the text are rare photographs and historic maps. **8.5" X 11", 92 ppgs. Retail Price: $9.99**

Geology of the Wallowa Mountains of Oregon: Part II (Volume 2) - Originally published in 1938, this important publication on Oregon Mining has not been available for nearly seventy five years. Included are details on the geology of this unique portion of North Eastern Oregon. This is the first part of a two book series on the area. Accompanying the text are rare photographs and historic maps. **8.5" X 11", 94 ppgs. Retail Price: $9.99**

Field Identification of Minerals For Oregon Prospectors - Originally published in 1940, this important publication on Oregon Mining has not been available for nearly seventy five years. Included in this volume is an easy system for testing and identifying a wide range of minerals that might be found by prospectors, geologists and rockhounds in the State of Oregon, as well as in other locales. Topics include how to put together your own field testing kit and how to conduct rudimentary tests in the field. This volume is written in a clear and concise way to make it useful even for beginners. **8.5" X 11", 158 ppgs. Retail Price: $14.99**

The Bohemia Mining District of Oregon - Originally published in 1900, this important publication on Oregon Mining has not been available for over a century. Included in this volume are important insights into the famous Bohemia Mining District of Oregon, including the histories and locations of important gold mines in the area such as the Ophir Mine, Clarence, Acturas, Peek-a-boo, White Swan, Combination Mine, the Musick Mine, The California, White Ghost, The Mystery, Wall Street, Vesuvius, Story, Lizzie Bullock, Delta, Elsie Dora, Golden Slipper, Broadway, Champion Mine, Knott, Noonday, Helena, White Wings, Riverside and others. Also included are notes on the nearby Blue River Mining District. **8.5" X 11", 58 ppgs. Retail Price: $9.99**

The Gold Fields of Eastern Oregon - Unavailable since 1900, this publication was originally compiled by the Baker City Chamber of Commerce Offering important insights into the gold mining history of Eastern Oregon, "The Gold Fields of Eastern Oregon" sheds a rare light on many of the gold mines that were operating at the turn of the 19th Century in Baker County and Grant County in North Eastern Oregon. Some of the areas featured include the Cable Cove District, Baisely-Elhorn, Granite, Red Boy, Bonanza, Susanville, Sparta, Virtue, Vaughn, Sumpter, Burnt River, Rye Valley and other mining districts. Included is basic information on not only many gold mines that are well known to those interested in Eastern Oregon mining history, but also many mines and prospects which have been mostly lost to the passage of time. Accompanying are numerous rare photos **8.5" X 11", 78 ppgs. Retail Price: $10.99**

Gold Mining in Eastern Oregon - Originally published in 1938, this important publication on Oregon Mining has not been available for over a century. Included in this volume are important insights into the famous mining districts of Eastern Oregon during the late 1930's. Particular attention is given to those gold mines with milling and concentrating facilities in the Greenhorn, Red Boy, Alamo, Bonanza, Granite, Cable Cove, Cracker Creek, Virtue, Keating, Medical Springs, Sanger, Sparta, Chicken Creek, Mormon Basin, Connor Creek, Cornucopia and the Bull Run Mining Districts. Some of the mines featured include the Ben Harrison, North Pole-Columbia, Highland Maxwell, Baisley-Elkhorn, White Swan, Balm Creek, Twin Baby, Gem of Sparta, New Deal, Gleason, Gifford-Johnson, Cornucopia, Record, Bull Run, Orion and others. Of particular interest are the mill flow sheets and descriptions of milling operations of these mines. **8.5" X 11", 68 ppgs. Retail Price: $8.99**

The Gold Belt of the Blue Mountains of Oregon - Originally published in 1901, this important publication on Oregon Mining has not been available for over a century. Included in this volume are rare insights into the gold deposits of the Blue Mountains of North East Oregon, including the history of their early discovery and early production. Extensive details are offered on this important mining area's mineralogy and economic geology, as well as insights into nearby gold placers, silver deposits and copper deposits. Featured are the Elkhorn and Rock Creek mining districts, the Pocahontas district, Auburn and Minersville districts, Sumpter and Cracker Creek, Cable Cove, the Camp Carson district, Granite, Alamo, Greenhorn, Robinsonville, the Upper Burnt River Valley and Bonanza districts, Susanville, Quartzburg, Canyon Creek, Virtue, the Copper Butte district, the North Powder River, Sparta, Eagle Creek, Cornucopia, Pine Creek, Lower Powder River, the Upper Snake River Canyon, Rye Valley, Lower Burnt River Valley, Mormon Basin, the Malheur and Clarks Creek districts, Sutton Creek and others. Of particular interest are important details on numerous gold mines and prospects in these mining districts, including their locations, histories, geology and other important information, as well as information on silver, copper and fire opal deposits. **8.5" X 11", 250 ppgs. Retail Price: $24.99**

Mining in the Cascades Range of Oregon - Originally published in 1938, this important publication on Oregon Mining has not been available for over seventy five years. Included in this volume are rare insights into the gold mines and other types of metal mines in the Cascades Mountain Range of Oregon. Some of the important mining areas covered include the famous Bohemia Mining District, the North Santiam Mining District, Quartzville Mining District, Blue River Mining District, Fall Creek Mining District, Oakridge District, Zinc District, Buzzard-Al Sarena District, Grand Cove, Climax District and Barron Mining District. Of particular interest are important details on over 100 mines and prospects in these mining districts, including their locations, histories, geology and other important information. 8.5" X 11", 170 ppgs. Retail Price: $14.99

Beach Gold Placers of the Oregon Coast - Originally published in 1934, this important publication on Oregon Mining has not been available for over 80 years. Included in this volume are rare insights into the beach gold deposits of the State of Oregon, including their locations, occurance, composition and geology. Of particular interest is information on placer platinum in Oregon's rich beach deposits. Also included are the locations and other information on some famous Oregon beach mines, including the Pioneer, Eagle, Chickamin, Iowa and beach placer mines north of the mouth of the Rogue River. 8.5" X 11", 60 ppgs. Retail Price: $8.99

Mineralogical Composition of the Sands of the Oregon Coast: From Coos Bay to the Columbia - Published in 1945, he text features hard to find information on the composition of the gold bearing black sands of the South West Oregon Coast, offering a unique insight to prospectors in search of Oregon's legendary beach gold. 104 ppgs, $9.99

Manganese Mining in Oregon - First released in 1942 and now out of print, this special reprint edition of "Manganese in Oregon" was originally published by the Oregon Department of Geology and Mineral Industries. The text features hard to find information on the mining of Manganese in Oregon, including details and maps of Oregon manganese mines and prospects. 108 ppgs, 9.99

Medford Oregon As A Mining Center - Written in 1912, this hard to find publication includes valuable insights into the mining history of South West Oregon. This small book contains interesting information on the gold, copper and mining industry in Southern Oregon as it existed just prior to World War One, shedding light on some of the important mines in the area. Included are rare photographs and vintage advertising of the day. 80 ppgs, 9.99

Mineral Resources of Curry County Oregon - First released in 1977 and now out of print, this special reprint edition of "Geology, Mineral Resources and Rock Materials of Curry County, Oregon" was originally published in cooperation of Curry County, Oregon and the Oregon Department of Geology and Mineral Industries. The text features hard to find information on not only the mining of gold and other metals in Curry County, but also aggregate mining in the area. 102 ppgs, 11.99

Origin of the Gold Bearing Black Sands of the Coast of South West Oregon - First released in 1943 and now out of print, this special reprint edition of "The Origin of the Black Sands of the South West Oregon Coast" was originally published by the Oregon Department of Geology and Mineral Industries. The text features hard to find information on the origin of the gold bearing black sands of the South West Oregon Coast, offering a unique insight to prospectors in search of Oregon's legendary beach gold. 52 ppgs, 8.99

South West Oregon Mining - Leading mining historian Kerby Jackson introduces us to six classic small mining publications on the Gold Mining Industry in Southern Oregon. This small book consists of a compilation of USGS J.S. Diller's "Mines of the Riddles Quadrangle", "The Rogue River Valley Coal Fields" and "Mineral Resources of the Grants Pass Quadrangle", the Grants Pass Commercial Club's rare publication "Mining in Josephine County, Oregon" and the USGS publication "The Distribution of Placer Gold in the Sixes River, South West Oregon". Also included is F.W. Libbey's legendary article on the Southern Oregon Mining Industry, "Lest We Forget", which appeared in the publication of the Oregon State Department of Geology and Mineral Industries in the early 1960's. This compilation offers a unique perspective on mining in South West Oregon and includes considerable information on mines in Josephine, Jackson and Coos Counties. 142 ppgs, 14.99

Geology and Mineral Resources of the Gasquet Quadrangle of California-Oregon - First published in 1953, it has been unavailable for over a century and sheds important light on the geological features and mineral resources of this portion of Northern California and Southern Oregon. 80 ppgs, 9.99

Idaho Mining Books

Gold in Idaho - Unavailable since the 1940's, this publication was originally compiled by the Idaho Bureau of Mines and includes details on gold mining in Idaho. Included is not only raw data on gold production in Idaho, but also valuable insight into where gold may be found in Idaho, as well as practical information on the gold bearing rocks and other geological features that will assist those looking for placer and lode gold in the State of Idaho. This volume also includes thirteen gold maps that greatly enhance the practical usability of the information contained in this small book detailing where to find gold in Idaho. **8.5" X 11", 72 ppgs. Retail Price: $9.99**

Geology of the Couer D'Alene Mining District of Idaho - Unavailable since 1961, this publication was originally compiled by the Idaho Bureau of Mines and Geology and includes details on the mining of gold, silver and other minerals in the famous Coeur D'Alene Mining District in Northern Idaho. Included are details on the early history of the Coeur D'Alene Mining District, local tectonic settings, ore deposit features, information on the mineral belts of the Osburn Fault, as well as detailed information on the famous Bunker Hill Mine, the Dayrock Mine, Galena Mine, Lucky Friday Mine and the infamous Sunshine Mine. This volume also includes sixteen hard to find maps. **8.5" X 11", 70 ppgs. Retail Price: $9.99**

The Gold Camps and Silver Cities of Idaho - Originally published in 1963, this important publication on Idaho Mining has not been available for nearly fifty years. Included are rare insights into the history of Idaho's Gold Rush, as well as the mad craze for silver in the Idaho Panhandle. Documented in fine detail are the early mining excitements at Boise Basin, at South Boise, in the Owyhees, at Deadwood, Long Valley, Stanley Basin and Robinson Bar, at Atlanta, on the famous Boise River, Volcano, Little Smokey, Banner, Boise Ridge, Hailey, Leesburg, Lemhi, Pearl, at South Mountain, Shoup and Ulysses, Yellow Jacket and Loon Creek. The story follows with the appearance of Chinese miners at the new mining camps on the Snake River, Black Pine, Yankee Fork, Bay Horse, Clayton, Heath, Seven Devils, Gibbonsville, Vienna and Sawtooth City. Also included are special sections on the Idaho Lead and Silver mines of the late 1800's, as well as the mining discoveries of the early 1900's that paved the way for Idaho's modern mining and mineral industry. Lavishly illustrated with rare historic photos, this volume provides a one of a kind documentary into Idaho's mining history that is sure to be enjoyed by not only modern miners and prospectors who still scour the hills in search of nature's treasures, but also those enjoy history and tromping through overgrown ghost towns and long abandoned mining camps. **8.5" X 11", 186 ppgs. Retail Price: $14.99**

Ore Deposits and Mining in North Western Custer County Idaho - Unavailable since 1913, this important publication was originally published by the Us Department of the Interior and has been unavailable for a century. Included are fine details on the geology, geography, gold placers and gold and silver bearing quartz veins of the mining region of North West Custer County, Idaho. Of particular interest is a rare look at the mines and prospects of the region, including those such as the Ramshorn Mine, SkyLark, Riverview, Excelsior, Beardsley, Pacific, Hoosier, Silver Brick, Forest Rose and dozens of others in the Bay Horse Mining District. Also covered are the mines of the Yankee Fork District such as the Lucky Boy, Badger, Black, Enterprise, Charles Dickens, Morrison, Golden Sunbeam, Montana, Golden Gate and others, as well as those in the Loon Mining District. **8.5" X 11", 126 ppgs. Retail Price: $12.99**

Gold Rush To Idaho - Unavailable since 1963, this important publication was originally published by the Idaho Bureau of Mines and has been unavailable for 50 years. "Gold Rush To Idaho" revisits the earliest years of the discovery of gold in Idaho Territory and introduces us to the conditions that the pioneer gold seekers met when they blazed a trail through the wilderness of Idaho's mountains and discovered the precious yellow metal at Oro Fino and Pierce. Subsequent rushes followed at places like Elk City, Newsome, Clearwater Station, Florence, Warrens and elsewhere. Of particular interest is a rare look at the hardships that the first miners in Idaho met with during their day to day existences and their attempts to bring law and order to their mining camps. **8.5" X 11", 88 ppgs. Retail Price: $9.99**

The Geology and Mines of Northern Idaho and North Western Montana - Unavailable since 1909, this important publication was originally published by the Us Department of the Interior and has been unavailable for a century. Included are fine details on the geology and geography of the mining regions of Northern Idaho and North Western Montana. Of particular interest is a rare look at the mines and prospects of the region, including those in the Pine Creek Mining District, Lake Pend Oreille district, Troy Mining District, Sylvanite District, Cabinet Mining District, Prospect Mining District and the Missoula Valley. Some of the mines featured include the Iron Mountain, Silver Butte, Snowshoe, Grouse Mountain Mine and others. **8.5" X 11", 142 ppgs. Retail Price: $12.99**

Mining in the Alturas Quadrangle of Blaine County Idaho - Unavailable since 1922, this important publication was originally published by the Idaho Bureau of Mines and has been unavailable for ninety years. Topics include the geology, rock formations and the formation of ore deposits in this important mining area of Idaho. Of particular focus is information on the local geology, quartz veins and ore deposits of this portion of Idaho. Included are hard to find details, including the descriptions and locations of numerous gold and silver mines in the area including the Silver King, Pilgrim, Columbia, Lone Jack, Sunbeam, Pride of the West, Lucky Boy, Scotia, Atlanta, Beaver-Bidwell and others mines and prospects. **8.5" X 11", 56 ppgs. Retail Price: $8.99**

<u>**Mining in Lemhi County Idaho**</u> - Originally published in 1913, this important book on Idaho Mining has not been available to miners for over a century. Included are rare insights into hundreds of gold, silver, copper and other mines in this famous Idaho mining area. Details include the locations, geology, history, production and other facts of the mines of this region, not only gold and silver hardrock mines, but also gold placer mines, lead-silver deposits, copper mines, cobalt-nickel deposits, tungsten and tin mines . It is lavishly illustrated with hard to find photos of the period and rare mining maps. Some of the vicinities featured include the Nicholia Mining District, Spring Mountain District, Texas District, Blue Wing District, Junction District, McDevitt District, Pratt Creek, Eldorado District, Kirtley Creek, Carmen Creek, Gibbonsville, Indian Creek, Mineral Hill District, Mackinaw, Eureka District, Blackbird District, YellowJacket District, Gravel Range District, Junction District, Parker Mountain and other mining districts. **8.5" X 11", 226 ppgs. Retail Price: $19.99**

<u>**Mining in Shoshone County Idaho**</u> - First published in 1923, it has been unavailable for over a century and sheds important light on the mining history of Shoshone County, Idaho. Some of the topics include the history of mining in Shoshone County, a look at the local geology and ore characteristics of lead-silver deposits, zinc deposits, copper, antimony, gold and other minerals. Also included are insights into the history, production, characteristics and locations of numerous mines in the area. 198 ppgs, 15.99

Utah Mining Books

<u>**Fluorite in Utah**</u> - Unavailable since 1954, this publication was originally compiled by the USGS, State of Utah and U.S. Atomic Energy Commission and details the mining of fluorspar, also known as fluorite in the State of Utah. Included are details on the geology and history of fluorspar (fluorite) mining in Utah, including details on where this unique gem mineral may be found in the State of Utah. **8.5" X 11", 60 ppgs. Retail Price: $8.99**

<u>**The Gold Hill Mining District of Utah**</u> - First published in 1935, it has been unavailable since those days and sheds important light on the mines, history and geology of Utah's Gold Hill Mining District. Included are rare insights into this important mining area, including the locations, histories and details of numerous mines. This volume is well illustrated with geological diagrams, as well as hard to find maps of some of the most important mines in this district. 202 ppgs., 19.99

<u>**The Mines, Miners and Minerals of Utah**</u> - First published in 1896, it has been unavailable since those days and sheds important light on the early mines and miners of Pioneer Utah, as well as the minerals which they won from the earth by laborious hard physical labor and sheer determination. Included are rare insights into the early mining history of Utah, as well details on hundreds of gold, silver and copper mines. 376 ppgs., 24.99

California Mining Books

<u>**The Tertiary Gravels of the Sierra Nevada of California**</u> - Mining historian Kerby Jackson introduces us to a classic mining work by Waldemar Lindgren in this important re-issue of The Tertiary Gravels of the Sierra Nevada of California. Unavailable since 1911, this publication includes details on the gold bearing ancient river channels of the famous Sierra Nevada region of California. **8.5" X 11", 282 ppgs. Retail Price: $19.99**

<u>**The Mother Lode Mining Region of California**</u> - Unavailable since 1900, this publication includes details on the gold mines of California's famous Mother Lode gold mining area. Included are details on the geology, history and important gold mines of the region, as well as insights into historic mining methods, mine timbering, mining machinery, mining bell signals and other details on how these mines operated. Also included are insights into the gold mines of the California Mother Lode that were in operation during the first sixty years of California's mining history. **8.5" X 11", 176 ppgs. Retail Price: $14.99**

<u>**Lode Gold of the Klamath Mountains of Northern California and South West Oregon**</u> - Unavailable since 1971, this publication was originally compiled by Preston E. Hotz and includes details on the lode mining districts of Oregon and California's Klamath Mountains. Included are details on the geology, history and important lode mines of the French Gulch, Deadwood, Whiskeytown, Shasta, Redding, Muletown, South Fork, Old Diggings, Dog Creek (Delta), Bully Choop (Indian Creek), Harrison Gulch, Hayfork, Minersville, Trinity Center, Canyon Creek, East Fork, New River, Denny, Liberty (Black Bear), Cecilville, Callahan, Yreka, Fort Jones and Happy Camp mining districts in California, as well as the Ashland, Rogue River, Applegate, Illinois River, Takilma, Greenback, Galice, Silver Peak, Myrtle Creek and Mule Creek districts of South Western Oregon. Also included are insights into the mineralization and other characteristics of this important mining region. **8.5" X 11", 100 ppgs. Retail Price: $10.99**

<u>**Mines and Mineral Resources of Shasta County, Siskiyou County, Trinity County: California**</u> - Unavailable since 1915, this publication was originally compiled by the California State Mining Bureau and includes details on the gold mines of this area of Northern California. Also included are insights into the mineralization and other characteristics of this important mining region, as well as the location of historic gold mines. **8.5" X 11", 204 ppgs. Retail Price: $19.99**

Geology of the Yreka Quadrangle, Siskiyou County, California - Unavailable since 1977, this publication was originally compiled by Preston E. Hotz and includes details on the geology of the Yreka Quadrangle of Siskiyou County, California. Also included are insights into the mineralization and other characteristics of this important mining region. **8.5" X 11", 78 ppgs. Retail Price: $7.99**

Mines of San Diego and Imperial Counties, California - Originally published in 1914, this important publication on California Mining has not been available for a century. This publication includes important information on the early gold mines of San Diego and Imperial County, which were some of the first gold fields mined in California by early Spanish and Mexican miners before the 49ers came on the scene. Included are not only details on early mining methods in the area, production statistics and geological information, but also the location of the early gold mines that helped make California "The Golden State". Also included are details on the mining of other minerals such as silver, lead, zinc, manganese, tungsten, vanadium, asbestos, barite, borax, cement, clay, dolomite, fluospar, gem stones, graphite, marble, salines, petroleum, stronium, talc and others. **8.5" X 11", 116 ppgs. Retail Price: $12.99**

Mines of Sierra County, California - Unavailable since 1920, this publication was originally compiled by the California State Mining Bureau and includes details on the gold mines of Sierra County, California. Also included are insights into the mineralization and other characteristics of this important mining region, as well as the location of historic gold mines. **8.5" X 11", 156 ppgs. Retail Price: $19.99**

Mines of Plumas County, California - Unavailable since 1918, this publication was originally compiled by the California State Mining Bureau and includes details on the gold mines of Plumas County, California. Also included are insights into the mineralization and other characteristics of this important mining region, as well as the location of historic gold mines. **8.5" X 11", 200 ppgs. Retail Price: $19.99**

Mines of El Dorado, Placer, Sacramento and Yuba Counties, California - Originally published in 1917, this important publication on California Mining has not been available for nearly a century. This publication includes important information on the early gold mines of El Dorado County, Placer County, Sacramento County and Yuba County, which were some of the first gold fields mined by the Forty-Niners during the California Gold Rush. Included are not only details on early mining methods in the area, production statistics and geological information, but also the location of the early gold mines that helped make California "The Golden State". Also included are insights into the early mining of chrome, copper and other minerals in this important mining area. **8.5" X 11", 204 ppgs. Retail Price: $19.99**

Mines of Los Angeles, Orange and Riverside Counties, California - Originally published in 1917, this important publication on California Mining has not been available for nearly a century. This publication includes important information on the early gold mines of Los Angeles County, Orange County and Riverside County, which were some of the first gold fields mined in California by early Spanish and Mexican miners before the 49ers came on the scene. Included are not only details on early mining methods in the area, production statistics and geological information, but also the location of the early gold mines that helped make California "The Golden State". **8.5" X 11", 146 ppgs. Retail Price: $12.99**

Mines of San Bernadino and Tulare Counties, California - Originally published in 1917, this important publication on California Mining has not been available for nearly a century. This publication includes important information on the early gold mines of San Bernadino and Tulare County, which were some of the first gold fields mined in California by early Spanish and Mexican miners before the 49ers came on the scene. Included are not only details on early mining methods in the area, production statistics and geological information, but also the location of the early gold mines that helped make California "The Golden State". Also included are details on the mining of other minerals such as copper, iron, lead, zinc, manganese, tungsten, vanadium, asbestos, barite, borax, cement, clay, dolomite, fluospar, gem stones, graphite, marble, salines, petroleum, stronium, talc and others. **8.5" X 11", 200 ppgs. Retail Price: $19.99**

Chromite Mining in The Klamath Mountains of California and Oregon - Unavailable since 1919, this publication was originally compiled by J.S. Diller of the United States Department of Geological Survey and includes details on the chromite mines of this area of Northern California and Southern Oregon. Also included are insights into the mineralization and other characteristics of this important mining region, as well as the location of historic mines. Also included are insights into chromite mining in Eastern Oregon and Montana. **8.5" X 11", 98 ppgs. Retail Price: $9.99**

Mines and Mining in Amador, Calaveras and Tuolumne Counties, California - Unavailable since 1915, this publication was originally compiled by William Tucker and includes details on the mines and mineral resources of this important California mining area. Included are details on the geology, history and important gold mines of the region, as well as insights into other local mineral resources such as asbestos, clay, copper, talc, limestone and others. Also included are insights into the mineralization and other characteristics of this important portion of California's Mother Lode mining region. **8.5" X 11", 198 ppgs. Retail Price: $14.99**

The Cerro Gordo Mining District of Inyo County California - Unavailable since 1963, this publication was originally compiled by the United States Department of Interior. Included are insights into the mineralization and other characteristics of this important mining region of Southern California. Topics include the mining of gold and silver in this important mining district in Inyo County, California, including details on the history, production and locations of the Cerro Gordo Mine, the Morning Star Mine, Estelle Tunnel, Charles Lease Tunnel, Ignacio, Hart, Crosscut Tunnel, Sunset, Upper Newtown, Newtown, Ella, Perseverance, Newsboy, Belmont and other silver and gold mines in the Cerro Gordo Mining District. This volume also includes important insights into the fossil record, geologic formations, faults and other aspects of economic geology in this California mining district. **8.5" X 11", 104 ppgs. Retail Price: $10.99**

Mining in Butte, Lassen, Modoc, Sutter and Tehama Counties of California - Unavailable since 1917, this publication was originally compiled by the United States Department of Interior. Included are insights into the mineralization and other characteristics of this important mining region of California. Topics include the mining of asbestos, chromite, gold, diamonds and manganese in Butte County, the mining of gold and copper in the Hayden Hill and Diamond Mountain mining districts of Lassen County, the mining of coal, salt, copper and gold in the High Grade and Winters mining districts of Modoc County, gold mining in Sutter County and the mining of gold, chromite, manganese and copper in Tehama County. This volume also includes the production records and locations of numerous mines in this important mining region. **8.5" X 11", 114 ppgs. Retail Price: $11.99**

Mines of Trinity County California - Originally published in 1965, this important publication on California Mining has not been available for nearly fifty years. This publication includes important information on mines and mining in Trinity County, California, as well insights into the mineralization and geology of this important mining area in Northern California. Included are extensive details on hardrock and placer gold mines and prospects, including charts showing the locations of these historic mines.. **8.5" X 11", 144 ppgs. Retail Price: $12.99**

Mines of Kern County California - Originally published in 1962, this important publication on California Mining has not been available for nearly fifty years. This publication includes important information on mines and mining in Kern County, California, as well insights into the mineralization and geology of this important mining area in California. Included are extensive details on hardrock and placer gold mines and prospects, including charts showing the locations of these historic mines. **8.5" X 11", 398 ppgs. Retail Price: $24.99**

Mines of Calaveras County California - Originally published in 1962, this important publication on California Mining has not been available for nearly fifty years. This publication includes important information on mines and mining in Calaveras County, California, as well insights into the mineralization and geology of this important mining area in Northern California. Included are extensive details on hardrock and placer gold mines and prospects, including charts showing the locations of these historic mines. **8.5" X 11", 236 ppgs. Retail Price: $19.99**

Lode Gold Mining in Grass Valley California - Unavailable since 1940, this publication was originally compiled by the United States Department of Interior. Included are insights into the gold mineralization and other characteristics of this important mining region of Nevada County, California. This volume also includes important insights into the geologic formations, faults and other aspects of economic geology in this California mining district. Of particular interest are the fine details on many hardrock gold mines in the area, including their locations, histories, development and mineralization. Some of the mines featured include the Gold Hill Mine, Massachusetts Hill, Boundary, Peabody, Golden Center, North Star, Omaha, Lone Jack, Homeward Bound, Hartery, Wisconsin, Allison Ranch, Phoenix, Kate Hayes, W.Y.O.D., Empire, Rich Hill, Daisy Hill, Orleans, Sultana, Centennial, Conlin, Ben Franklin, Crown Point and many others. **8.5" X 11", 148 ppgs. Retail Price: $12.99**

Lode Mining in the Alleghany District of Sierra County California - Unavailable since 1913, this publication was originally compiled by the United States Department of Interior. Included are insights into the mineralization and other characteristics of this important mining region of Sierra County. Included are details on the history, production and locations of numerous hardrock gold mines in this famous California area, including the Tightner Mine, Minnie D., Osceola, Eldorado, Twenty One, Sherman, Kenton, Oriental, Rainbow, Plumbago, Irelan, Gold Canyon, North Fork, Federal, Kate Hardy and others. This volume also includes important insights into the fossil record, geologic formations, faults and other aspects of economic geology in this California mining district. **8.5" X 11", 48 ppgs. Retail Price: $7.99**

Six Months In The Gold Mines During The California Gold Rush - Unavailable since 1850, this important work is a first hand account of one "49'ers" personal experience during the great California Gold Rush, shedding important light on one of the most exciting periods in the history of not only California, but also the world. Compiled from journals written between 1847 and 1849 by E. Gould Buffum, a native of New York, "Six Months In The Gold Mines During The California Gold Rush" offers a rare look into the day to day lives of the people who came to California to work in her gold mines when the state was still a great frontier. **8.5" X 11", 290 ppgs. Retail Price: $19.99**

<u>Quartz Mines of the Grass Valley Mining District of California</u> - Unavailable since 1867, this important publication has not been available since those days. This rare publication offers a short dissertation on the early hardrock mines in this important mining district in the California Mother Lode region between the 1850's and 1860's. Also included are hard to find details on the mineralization and locations of these mines, as well as how they were operated in those day. **8.5" X 11", 44 ppgs. Retail Price: $8.99**

<u>Gold Rush on the Feather River</u> - First published in 1924, this short publication by G.C. Mansfield sheds important light on the early history of gold mining on the Feather River. Included are rare insights into the first decade of gold mining and the early mining camps of the Feather River during the 1850's. 64 ppgs., 9.99

<u>The Bodie Mining District of California</u> - First published in 1986, it has been unavailable since those days and sheds important light on this famous mining area. Included are the history, characteristics and locations of numerous old mines around the ghost town of Bodie.
64 ppgs, 8.99

<u>Geology and Mineral Resources of the Gasquet Quadrangle of California-Oregon</u> - First published in 1953, it has been unavailable for over a century and sheds important light on the geological features and mineral resources of this portion of Northern California and Southern Oregon.
80 ppgs, 9.99

Alaska Mining Books

<u>Ore Deposits of the Willow Creek Mining District, Alaska</u> - Unavailable since 1954, this hard to find publication includes valuable insights into the Willow Creek Mining District near Hatcher Pass in Alaska. The publication includes insights into the history, geology and locations of the well known mines in the area, including the Gold Cord, Independence, Fern, Mabel, Lonesome, Snowbird, Schroff-O'Neil, High Grade, Marion Twin, Thorpe, Webfoot, Kelly-Willow, Lane, Holland and others. **8.5" X 11", 96 ppgs. Retail Price: $9.99**

<u>The Juneau Gold Belt of Alaska</u> - Unavailable since 1906, this hard to find publication includes valuable insights into the gold mines around Juneau, Alaska. The publication includes important details into the history, geology and locations of the well known gold mines and prospects in the area, including those around Windham Bay, Holkham Bay, Port Snettisham, on Grindstone and Rhine Creeks, Gold Creek, Douglas Island, Salmon Creek, Lemon Creek, Nugget Creek, from the Mendenhall River to Berners Bay, McGinnis Creek, Montana Creek, Peterson Creek, Windfall Creek, the Eagle River, Yankee Basin, Yankee Curve, Kowee Creek and elsewhere. Not only are gold placer mines included, but also hardrock gold mines. **8.5" X 11", 224 ppgs. Retail Price: $19.99**

<u>Mining in the Jumbo Basin of Alaska</u> - Unavailable since 1953, this hard to find publication includes valuable insights into the mines and geology of the Jumbo Basin. The publication includes important details into the history, geology and locations of the well known gold mines and prospects in the famous Jumbo Basin Mining Region of Alaska.
72 ppgs, 9.99

<u>The Rampart Placer Gold Region of Alaska</u> - Unavailable since 1906, this hard to find publication includes valuable insights into the placer gold mines of the Rampart Mining Region. The publication includes important details into the history, geology and locations of the well known gold mines and prospects in the famous Rampart Mining Region of Alaska.
78 ppgs, 10.99

Arizona Mining Books

<u>Mines and Mining in Northern Yuma County Arizona</u> - Originally published in 1911, this important publication on Arizona Mining has not been available for over a hundred years. Included are rare insights into the gold, silver, copper and quicksilver mines of Yuma County, Arizona together with hard to find maps and photographs. Some of the mines and mining districts featured include the Planet Copper Mine, Mineral Hill, the Clara Consolidated Mine, Viati Mine, Copper Basin prospect, Bowman Mine, Quartz King, Billy Mack, Carnation, the Wardwell and Osbourne, Valensuella Copper, the Mariquita, Colonial Mine, the French American, the New York-Plomosa, Guadalupe, Lead Camp, Mudersbach Copper Camp, Yellow Bird, the Arizona Northern (Salome Strike), Bonanza (Harqua Hala), Golden Eagle, Hercules, Socorro and others. **8.5" X 11", 144 ppgs. Retail Price: $11.99**

<u>The Aravaipa and Stanley Mining Districts of Graham County Arizona</u> - Originally published in 1925, this important publication on Arizona Mining has not been available for nearly ninety years. Included are rare insights into the gold and silver mines of these two important mining districts, together with hard to find maps. **8.5" X 11", 140 ppgs. Retail Price: $11.99**

Gold in the Gold Basin and Lost Basin Mining Districts of Mohave County, Arizona - This volume contains rare insights into the geology and gold mineralization of the Gold Basin and Lost Basin Mining Districts of Mohave County, Arizona that will be of benefit to miners and prospectors. Also included is a significant body of information on the gold mines and prospects of this portion of Arizona. This volume is lavishly illustrated with rare photos and mining maps. 8.5" X 11", 188 ppgs. Retail Price: $19.99

Mines of the Jerome and Bradshaw Mountains of Arizona - This important publication on Arizona Mining has not been available for ninety years. This volume contains rare insights into the geology and ore deposits of the Jerome and Bradshaw Mountains of Arizona that will be of benefit to miners and prospectors who work those areas. Included is a significant body of information on the mines and prospects of the Verde, Black Hills, Cherry Creek, Prescott, Walker, Groom Creek, Hassayampa, Bigbug, Turkey Creek, Agua Fria, Black Canyon, Peck, Tiger, Pine Grove, Bradshaw, Tintop, Humbug and Castle Creek Mining Districts. This volume is lavishly illustrated with rare photos and mining maps. 8.5" X 11", 218 ppgs. Retail Price: $19.99

The Ajo Mining District of Pima County Arizona - This important publication on Arizona Mining has not been available for nearly seventy years. This volume contains rare insights into the geology and mineralization of the Ajo Mining District in Pima County, Arizona and in particular the famous New Cornelia Mine. 8.5" X 11", 126 ppgs. Retail Price: $11.99

Mining in the Santa Rita and Patagonia Mountains of Arizona - Originally published in 1915, this important publication on Arizona Mining has not been available for nearly a century. Included are rare insights into hundreds of gold, silver, copper and other mines in this famous Arizona mining area. Details include the locations, geology, history, production and other facts of the mines of this region. 8.5" X 11", 394 ppgs. Retail Price: $24.99

Mining in the Bisbee Quadrangle of Arizona - Originally published in 1906, this important publication on Arizona Mining has not been available for nearly a century. Included are rare insights into hundreds of gold, silver, copper and other mines in this famous Arizona mining area. Details include the locations, geology, history, production and other facts of the mines of this important mining region. 8.5" X 11", 188 ppgs. Retail Price: $14.99

Placer Gold Mining in Arizona - Unavailable since 1922, this hard to find publication includes valuable insights into the placer gold mines of the Arizona. Originally released as "Placer Gold of Arizona", despite its small size, this publication includes important details into the history, geology and locations of the well known placer gold mines and prospects in the State of Arizona. 48 ppgs, 8.99

Gold and Copper Mining near Payson, Arizona - Written in 1915, this hard to find publication includes valuable insights into the gold and copper mining industry of Arizona. Highlighted here are the gold and copper mines near Payson, Arizona. 68 ppgs, 8.99

Lode Gold Mining in Arizona - Unavailable since 1934, this hard to find publication, originally released as "Arizona Lode Gold Mines and Gold Mining" includes valuable insights into the gold mining industry of Arizona. Included are valuable insights into over 150 hardrock gold mines in over 30 different mining districts in Arizona. 278 ppgs, 21.99

Mining in the Dragoon Quadrangle of Cochise County, Arizona - Unavailable since 1964, this hard to find publication includes valuable insights into the mines of the Dragoon Quadrangle Mining Region. The publication includes important details into the history, geology and locations of the well known mines and prospects in this famous mining region of Arizona. 224 ppgs., 19.99

Directory of Operating Mines in Arizona in 1915 - Unavailable since 1916, this hard to find publication includes valuable insights into the mines of Arizona. This small publication includes a complete list of the mines that were operating in the State of Arizona during 1915 and includes details such as general location, owners and some basic facts about each mining operation.52 ppgs. 8.99

Arizona Ore Deposits - Unavailable since 1938, this hard to find publication includes valuable insights into some ore deposits of Arizona. Included are valuable insights into the formation and characteristics of valuable ore deposits in the Jerome, Miami, Inspiration, Clifton, Morenci, Ray, Ajo, Eureka, Tombstone and Magma mining districts. Included are details into some of the major gold, silver and copper mines of these important Arizona mining areas. 160 ppgs, 14.99

Montana Mining Books

A History of Butte Montana: The World's Greatest Mining Camp - First published in 1900 by H.C. Freeman, this important publication sheds a bright light on one of the most important mining areas in the history of The West. Together with his insights, as well as rare photographs of the periods, Harry Freeman describes Butte and its vicinity from its early beginnings, right up to its flush years when copper flowed from its mines like a river. At the time of publication, Butte, Montana was known worldwide as "The Richest Mining Spot On Earth" and produced not only vast amounts of copper, but also silver, gold and other metals from its mines. Freeman illustrates, with great detail, the most important mines in the vicinity of Butte, providing rare details on their owners, their history and most importantly, how the mines operated and how their treasures were extracted. Of particular interest are the dozens of rare photographs that depict mines such as the famous Anaconda, the Silver Bow, the Smoke House, Moose, Paulin, Buffalo, Little Minah, the Mountain Consolidated, West Greyrock, Cora, the Green Mountain, Diamond, Bell, Parnell, the Neversweat, Nipper, Original and many others. **8.5" X 11", 142 ppgs. Retail Price: $12.99**

The Butte Mining District of Montana - This important publication on Montana Mining has not been available for over a century. Included are rare insights into the gold, copper and silver mines of Butte, Montana together with hard to find maps and photographs. Some of the topics include the early history of gold, silver and copper mining in the Butte area, insight into the geology of its mining areas, the local distribution of gold, silver and copper ores, as well their composition and how to identify them. Also included are detailed facts about the mines in the Butte Mining District, including the famous Anaconda Mine, Gagnon, Parrot, Blue Vein, Moscow, Poulin, Stella, Buffalo, Green Mountain, Wake Up Jim, the Diamond-Bell Group, Mountain Consolidated, East Greyrock, West Greyrock, Snowball, Corra, Speculator, Adirondack, Miners Union, the Jessie-Edith May Group, Otisco, Iduna, Colorado, Lizzie, Cambers, Anderson, Hesperus, Preferencia and dozens of others. **8.5" X 11", 298 ppgs. Retail Price: $24.99**

Mines of the Helena Mining Region of Montana - This important publication on Montana Mining has not been available for over a century. Included are rare insights into the gold, copper and silver mines of the vicinity of Helena, Montana, including the Marysville Mining District, Elliston Mining District, Rimini Mining District, Helena Mining District, Clancy Mining District, Wickes Mining District, Boulder and Basin Mining Districts and the Elkhorn Mining District. Some of the topics include the early history of gold, silver and copper mining in the Helena area, insight into the geology of its mining areas, the local distribution of gold, silver and copper ores, as well their composition and how to identify them. Also included are detailed facts, history, geology and locations of over one hundred gold, silver and copper mines in the area . **8.5" X 11", 162 ppgs, Retail Price: $14.99**

Mines and Geology of the Garnet Range of Montana - This important publication on Montana Mining has not been available for over a century. Included are rare insights into the gold, copper and silver mines of the vicinity of this important mining area of Montana. Some of the topics include the early history of gold, silver and copper mining in the Garnet Mountains, insight into the geology of its mining areas, the local distribution of gold, silver and copper ores, as well their composition and how to identify them. Also included are detailed facts, history, geology and locations of numerous gold, silver and copper mines in the area . **8.5" X 11", 100 ppgs, Retail Price: $11.99**

Mines and Geology of the Philipsburg Quadrangle of Montana - This important publication on Montana Mining has not been available for over a century. Included are rare insights into the gold, copper and silver mines of the vicinity of this important mining area of Montana. Some of the topics include the early history of gold, silver and copper mining in the Philipsburg Quadrangle, insight into the geology of its mining areas, the local distribution of gold, silver and copper ores, as well their composition and how to identify them. Also included are detailed facts, history, geology and locations of over one hundred gold, silver and copper mines in the area **8.5" X 11", 290 ppgs, Retail Price: $24.99**

Geology of the Marysville Mining District of Montana - Included are rare insights into the mining geology of the Marysville Mining District. Some of the topics include the early history of gold, silver and copper mining in the area, insight into the geology of its mining areas, the local distribution of gold, silver and copper ores, as well their composition and how to identify them. Also included are detailed facts, history, geology and locations of gold, silver and copper mines in the area **8.5" X 11", 198 ppgs, Retail Price: $19.99**

The Geology and Mines of Northern Idaho and North Western Montana - See listing under Idaho.

The History of Gold Dredging in Montana - Unavailable since 1916, this important publication was originally published by the Us Bureau of Mines and has been unavailable for a century. A century and more ago, giant dredging machines dug in Montana's rivers and creeks in search of illusive golden riches. First appearing in California in the 1850's, gold dredges finally reached their peak of development in Siberia and New Zealand before becoming popular again in the United States. This book offers a unique historical perspective on the gold dredges that once operated in Montana. This book on Montana mining history is lavishly illustrated with dozens of rare historic photos gold dredges that once operated in Montana, as well as hard to locate plans on how these dredges were designed. 120 ppgs., 11.99

Nevada Mining Books

The Bull Frog Mining District of Nevada - Unavailable since 1910, this publication was originally compiled by the United States Department of Interior. This volume also includes important insights into the geologic formations, faults and other aspects of economic geology in this Nevada mining district. Of particular interest are the fine details on many mines in the area, including their locations, histories, development and mineralization. Some of the mines featured include the National Bank Mine, Providence, Gibraltor, Tramps, Denver, Original Bullfrog, Gold Bar, Mayflower, Homestake-King and other mines and prospects. **8.5" X 11", 152 ppgs, Retail Price: $14.99**

History of the Comstock Lode - Unavailable since 1876, this publication was originally released by John Wiley & Sons. This volume also includes important insights into the famous Comstock Lode of Nevada that represented the first major silver discovery in the United States. During its spectacular run, the Comstock produced over 192 million ounces of silver and 8.2 million ounces of gold. Not only did the Comstock result in one of the largest mining rushes in history and yield immense fortunes for its owners, but it made important contributions to the development of the State of Nevada, as well as neighboring California. Included here are important details on not only the early development and history of the Comstock, but also rare early insight into its mines, ore and its geology.8.5" X 11", 244 ppgs, Retail Price: $19.99

The Pioche Mining District of Nevada - First published in 1932, it has been unavailable for over a century and sheds important light on the mining history of Nevada. Some of the topics include the history of mining in this district, as well as the characteristics of its mineral and ore deposits. Also included are insights into the history, production, characteristics and locations of numerous mines in the area. Some of the mines include the Combined Metals, Pioche, Ely Valley, No. 10, Poorman, Wide Awake, Alps, Prince, Virginia Louise, Half Moon, Abe Lincoln, Fairview, Bristol Silver, National, Vesuvius, Inman, Tempest, Hillside, Jackrabbit, Lucky Star, Fortuna, Mendha, Manhattan, Hamburg, Comet, Lyndon and others. 108 ppgs 10.99

The Yerington Mining District of Nevada - First published in 1932, it has been unavailable for over a century and sheds important light on the mining history of Nevada. Some of the topics include the history of mining in this district, as well as the characteristics of its mineral and ore deposits. Also included are insights into the history, production, characteristics and locations of numerous mines in the area. Some of the mines include the Bluestone, Mason Valley, Malachite, McConnell, Greenwood, Western Nevada, Ludwig, Douglas Hill, Casting Copper, Montana-Yerington, Empire, Jim Beatty, Terry and McFarland, Blue Jay and others. 92 ppgs, 10.99

The Genesis of the Ores of Tonopah Nevada - Unavailable since 1918, this hard to find publication includes valuable insights into the gold mines around Tonopah, Nevada. The publication includes important details into the geology of mines in the Tonopah Mining District of Nevada. 90 ppgs, 10.99

Mining Camps of Elko, Lander and Eureka Counties Nevada - Unavailable since 1910, this hard to find publication includes valuable insights into the mining camps of Elko, Lander and Eureka Counties, Nevada. The publication includes important details into the history of mines and mining in these three Nevada counties. 154 ppgs, 12.99

Ore Deposits of the Bullfrog Quadrangle - Unavailable since 1964 and released as "Geology of Bullfrog Quadrangle and Ore Deposits Related to Bullfrog Hills Caldera, Nye County, Nevada and Inyo County, California". The publication includes important details into the geology of mines in the Bullfrog Quadrangle of Nye County, Nevada and Inyo County, California. 52 ppgs, 9.99

Mining in Eureka County Nevada - Unavailable since 1879, this hard to find publication includes valuable insights into the early mining history off Eureka County, Nevada. The publication includes important details into the early history of the mines of Eureka County, as well as their development, production and how their ores were treated. Also included are details on the 1872 Mining Act, as well as the local rules, regulations and customs of the miners in Eureka County.134 ppgs, 12.99

Colorado Mining Books

Ores of The Leadville Mining District - Unavailable since 1926, this publication was originally compiled by the United States Department of Interior. This volume also includes important insights into the ores and mineralization of the Leadville Mining District in Colorado. Topics include historic ore prospecting methods, local geology, insights into ore veins and stockworks, the local trend and distribution of ore channels, reverse faults, shattered rock above replacement ore bodies, mineral enrichment in oxidized and sulphide zones and more. **8.5" X 11", 66 ppgs, Retail Price: $8.99**

Mining in Colorado - Unavailable since 1926, this publication was originally compiled by the United States Department of Interior. This volume also includes important insights into the mining history of Colorado from its early beginnings in the 1850's right up to the mid 1920's. Not only is Colorado's gold mining heritage included, but also its silver, copper, lead and zinc mining industry. Each mining area is treated separately, detailing the development of Colorado's mines on a county by county basis. **8.5" X 11", 284 ppgs, Retail Price: $19.99**

Gold Mining in Gilpin County Colorado - Unavailable since 1876, this publication was originally compiled by the Register Steam Printing House of Central City, Colorado. A rare glimpse at the gold mining history and early mines of Gilpin County, Colorado from their first discovery in the 1850's up to the "flush years" of the mid 1870's. Of particular interest is the history of the discovery of gold in Gilpin County and details about the men who made those first strikes. Special focus is given to the early gold mines and first mining districts of the area, many of which are not detailed in other books on Colorado's gold mining history. **8.5" X 11", 156 ppgs, Retail Price: $12.99**

Mining in the Gold Brick Mining District of Colorado - Important insights into the history of the Gold Brick Mining District, as well as its local geography and economic geology. Also included are the histories and locations of historic mines in this important Colorado Mining District, including the Cortland, Carter, Raymond, Gold Links, Sacramento, Bassick, Sandy Hook, Chronicle, Grand Prize, Chloride, Granite Mountain, Lucille, Gray Mountain, Hilltop, Maggie Mitchell, Silver Islet, Revenue, Roosevelt, Carbonate King and others. In addition to hardrock mining, are also included are details on gold placer mining in this portion of Colorado. **8.5" X 11", 140 ppgs, Retail Price: $12.99**

Ore Deposits of the London Fault of Colorado - First published in 1941, it has been unavailable since those days and sheds important light on the mines and mineral deposits of the London Fault in Central Colorado's Alma Mining District. This publication sheds important light on the gold veins and lead-silver deposits of the Alma Mining District. Included are geologic details on the London Mine, American Mine, Havigorst Tunnel, Ophir Mine, Mosher Tunnel, London-Butte Mine, Venture Shaft, Hard-To-Beat Mine, Oliver Twist Tunnel, Sacramento Mine, Mudsill Mine, Sherwood Mine, Wagner, Barcoe Tunnel and other mines in this important mining region. 110 ppgs., 10.99

The Mines of Colorado - First published in 1867, it has been unavailable since those days and sheds important light on Colorado's early mining history. Written shortly after the events took place, this publication sheds important light on the Pike's Peak Gold Rush, the discovery of gold on Ralston Creek and Dry Creek in the 1850's, as well as details on the first wave of miners into Colorado and their trials and tribulations as they crossed the Great Plains. Also included are details on early discoveries of lode gold in the mountainous regions of Colorado, details on the early mines hardrock and placer mines, and much more. It is a veritable treasure trove on Colorado's early mining history and will be of great importance to anyone who is interested in the mining of gold or other minerals in Colorado, as well as those interested in the history of the state. 478 ppgs., 29.99

The La Plata Mining District of Colorado - Originally titled "Geology and Ore Deposits in the Vicinity of the La Plata District of Colorado" and first published in 1949, it has been unavailable since those days and sheds important light on the mines and mineral deposits of the La Plata Mining District of Colorado. 214 ppgs., 19.99

Washington Mining Books

The Republic Mining District of Washington - Unavailable since 1910, this important publication was originally published by the Washington Geologic Survey and has been unavailable for a century. Topics include the geology, rock formations and the formation of ore deposits in this important mining area of Washington State. Also included are hard to find details on the geology, history and locations of dozens of mines in the area. Some of the mines featured include the New Republic Mine, Ben Hur, Morning Glory, the South Republic Mine, Quilp, Surprise, Black Tail, Lone Pine, San Poil, Mountain Lion, Tom Thumb, Elcaliph and many others. 8.5" X 11", 94 ppgs, **Retail Price: $10.99**

The Myers Creek and Nighthawk Mining Districts of Washington - Unavailable since 1911, this important publication was originally published by the Washington Geologic Survey and has been unavailable for a century. Topics include the geology, rock formations and the formation of ore deposits in these important mining areas of Washington State. Also included are hard to find details on the geology, history and locations of dozens of mines in the area. Some of the mines featured include the Grant Mine, Monterey, Nip and Tuck, Myers Creek, Number Nine, Neutral, Rainbow, Aztec, Crystal Butte, Apex, Butcher Boy, Molson, Mad River, Olentangy, Delate, Kelsey, Golden Chariot, Okanogan, Ohio, Forty-Ninth Parallel, Nighthawk, Favorite, Little Chopaka, Summit, Number One, California, Peerless, Caaba, Prize Group, Ruby, Mountain Sheep, Golden Zone, Rich Bar, Similkameen, Kimberly, Triune, Hiawatha, Trinity, Hornsilver, Maquae, Bellevue, Bullfrog, Palmer Lake, Ivanhoe, Copper World and many others. 8.5" X 11", 136 ppgs, **Retail Price: $12.99**

The Blewett Mining District of Washington - Unavailable since 1911, this important publication was originally published by the Washington Geologic Survey and has been unavailable for a century. Topics include the geology, rock formations and the formation of ore deposits in this important mining area of Washington State. Also included are hard to find details on the geology, history and locations of dozens of mines in the area. Some of the mines featured include the Washington Meteor, Alta Vista, Pole Pick, Blinn, North Star, Golden Eagle, Tip Top, Wilder, Golden Guinea, Lucky Queen, Blue Bell, Prospect, Homestake, Lone Rock, Johnson, and others. 8.5" X 11", 134 ppgs, **Retail Price: $12.99**

Silver Mining In Washington - Unavailable since 1955, this important publication was originally published by the Washington Geologic Survey. Featured are the hard to find locations and details pertaining to Washington's silver mines. 8.5" X 11", 180 ppgs, **Retail Price: $15.99**

The Mines of Snohomish County Washington - Unavailable since 1942, this important publication was originally published by the Washington Geologic Survey and has been unavailable for seventy years. Featured are details on a large number of gold, silver, copper, lead and other metallic mineral mines. Included are the locations of each historic mine, along with information on the commodity produced. 8.5" X 11", 98 ppgs, **Retail Price: $10.99**

The Mines of Chelan County Washington - Unavailable since 1943, this important publication was originally published by the Washington Geologic Survey and has been unavailable for seventy years. Featured are details on a large number of gold, silver, copper, lead and other metallic mineral mines. Included are the locations of each historic mine, along with information on the commodity. 8.5" X 11", 88 ppgs, **Retail Price: $9.99**

Metal Mines of Washington - Unavailable since 1921, this important publication was originally published by the Washington Geologic Survey and has been unavailable for nearly ninety years. Widely considered a masterpiece on the Washington Mining Industry, "Metal Mines of Washington" sheds light on the important details of Washington's early mining years. Featured are details on hundreds of gold, silver, copper, lead and other metallic mineral mines. Included are hard to find details on the mineral resources of this state, as well as the locations of historic mines. Lavishly illustrated with maps and historic photos and complete with a glossary to explain any technical terms found in the text, this is one of the most important works on mining in the State of Washington. No prospector or miner should be without it if they are interested in mining in Washington. 8.5" X 11", 396 ppgs, **Retail Price: $24.99**

Gem Stones In Washington - Unavailable since 1949, this important publication was originally published by the Washington Geologic Survey and has been unavailable since first published. Included are details on where to find naturally occurring gem stones in the State of Washington, including quartz crystal, amethyst, smoky quartz, milky quartz, agates, bloodstone, carnelian, chert, flint, jasper, onyx, petrified wood, opal, fire opal, hyalite and others. 8.5" X 11", 54 ppgs, **Retail Price: $8.99**

The Covada Mining District of Washington - Unavailable since 1913, this important publication was originally published by the Washington Geologic Survey and has been unavailable for a century. Topics include the geology, rock formations and the formation of ore deposits in this important mining area of Washington State. Also included are hard to find details on the geology, history and locations of dozens of mines in the area. Some of the mines featured include the Admiral, Advance, Algonkian, Big Bug, Big Chief, Big Joker, Black Hawk, Black Tail, Black Thorn, Captain, Cherokee Strip, Colorado, Dan Patch, Dead Shot, Etta, Good Ore, Greasy Run, Great Scott, Idora, IXL, Jay Bird, Kentucky Bell, King Solomon, Laurel, Laura S, Little Jay, Meteor, Neglected, Northern Light, Old Nell, Plymouth Rock, Polaris, Quandary, Reserve, Shoo Fly, Silver Plume, Three Pines, Vernie, White Rose and dozens of others. 8.5" X 11", 114 ppgs, **Retail Price: $10.99**

The Index Mining District of Washington - Unavailable since 1912, this important publication was originally published by the Washington Geologic Survey and has been unavailable for a century. Topics include the geology, rock formations and the formation of ore deposits in this important mining area of Washington State. Also included are hard to find details on the geology, history and locations of dozens of mines in the area. Some of the mines featured include the Sunset, Non-Pareil, Ethel Consolidated, Kittaning, Merchant, Homestead, Co-operative, Lost Creek, Uncle Sam, Calumet, Florence-Rae, Bitter Creek, Index Peacock, Gunn Peak, Helena, North Star, Buckeye. Copper Bell, Red Cross and others. **8.5" X 11", 114 ppgs, Retail Price: $11.99**

Mining & Mineral Resources of Stevens County Washington - Unavailable since 1920, this important publication was originally published by the Washington Geologic Survey and has been unavailable for a century. Topics include the geology, rock formations and the formation of ore deposits in these important mining areas of Washington State. Also included are hard to find details on the geology, history and locations of hundreds of mines in the area. **8.5" X 11", 372 ppgs, Retail Price: $24.99**

The Mines and Geology of the Loomis Quadrangle Okanogan County, Washington - Unavailable since 1972, this important publication was originally published by the Washington Geologic Survey and has been unavailable for a century. Topics include the geology, rock formations and the formation of ore deposits in this important mining area of Washington State. Also included are hard to find details on the geology, history and locations of dozens of gold, copper, silver and other mines in the area. **8.5" X 11", 150 ppgs, Retail Price: $12.99**

The Conconully Mining District of Okanogan County Washington - Unavailable since 1973, this important publication was originally published by the Washington Geologic Survey and has been unavailable for a century. Topics include the geology, rock formations and the formation of ore deposits in this important mining area of Washington State, which also includes Salmon Creek, Blue Lake and Galena. Also included are hard to find details on the geology, mining history and locations of dozens of mines in the area. Some of the mines include Arlington, Fourth of July, Sonny Boy, First Thought, Last Chance, War Eagle-Peacock, Wheeler, Mohawk, Lone Star, Woo Loo Moo Loo, Keystone, Hughes, Plant-Callahan, Johnny Boy, Leuena, Gubser, John Arthur, Tough Nut, Homestake, Key and many others **8.5" X 11", 68 ppgs, Retail Price: $8.99**

Wyoming Mining Books

Mining in the Laramie Basin of Wyoming - Unavailable since 1909, this publication was originally compiled by the United States Department of Interior. Also included are insights into the mineralization and other characteristics of this important mining region, especially in regards to coal, limestone, gypsum, bentonite clay, cement, sand, clay and copper. **8.5" X 11", 104 ppgs, Retail Price: $11.99**

New Mexico Mining Books

The Mogollon Mining District of New Mexico - Unavailable since 1927, this important publication was originally published by the US Department of Interior and has been unavailable for 80 years. Topics include the geology, rock formations and the formation of ore deposits in this important mining area in New Mexico. Of particular focus is information on the history and production of the ore deposits in this area, their form and structure, vein filling, their paragenesis, origins and ore shoots, as well as oxidation and supergene enrichment. Also included are hard to find details, including the descriptions and locations of numerous gold, silver and other types of mines, including the Eureka, Pacific, South Alpine, Great Western, Enterprise, Buffalo, Mountain View, Floride, Gold Dust, Last Chance, Deadwood, Confidence, Maud S., Deep Down, Little Fanney, Trilby, Johnson, Alberta, Comet, Golden Eagle, Cooney, Queen, the Iron Crown, Eberle, Clifton, Andrew Jackson mine, Mascot and others. **8.5" X 11", 144 ppgs, Retail Price: $12.99**

The Percha Mining District of Kingston New Mexico - Unavailable since 1883, this important publication was originally published by the Kingston Tribune and has been unavailable for over one hundred and thirty five years. Having been written during the earliest years of gold and silver mining in the Percha Mining District, unlike other books on the subject, this work offers the unique perspective of having actually been written while the early mining history of this area was still being made. In fact, the work was written so early in the development of this area that many of the notable mines in the Percha District were less than a few years old and were still being operated by their original discoverers with the same enthusiasm as when they were first located. Included are hard to find details on the very earliest gold and silver mines of this important mining district near Kingston in Sierra County, New Mexico. **8.5" X 11", 68 ppgs, Retail Price: $9.99**

East Coast Mining Books

<u>The Gold Fields of the Southern Appalachians</u> - Unavailable since 1895, this important publication was originally published by the US Department of Interior and has been unavailable for nearly 120 years. Topics include the geology, rock formations and the formation of ore deposits in this important mining area of the American South. Of particular focus is information on the history and statistics of the ore deposits in this area, their form and structure and veins. Also included are details on the placer gold deposits of the region. The gold fields of the Georgian Belt, Carolinian Belt and the South Mountain Mining District of North Carolina are all treated in descriptive detail. Included are hard to find details, including the descriptions and locations of numerous gold mines in Georgia, North Carolina and elsewhere in the American South. Also included are details on the gold belts of the British Maritime Provinces and the Green Mountains. **8.5" X 11", 104 ppgs, Retail Price: $9.99**

Gold Rush Tales Series

Millions in Siskiyou County Gold - In this first volume of the "Gold Rush Tales" series, leading mining historian and editor Kerby Jackson, introduces us to the story of how millions of dollars worth of gold was discovered in Siskiyou County during the California Gold Rush. Lavishly illustrated with photos from the 19th Century, this hard to find information was first published in 1897 and sheds important light onto the gold rush era in Siskiyou County, California and the experiences of the men who dug for the gold and actually found it. **8.5" X 11", 82 ppgs, Retail Price: $9.99**

The California Rand in the Days of '49 - In this second volume of the "Gold Rush Tales" series, leading mining historian and editor Kerby Jackson, introduces us to four tales from the California Gold Rush. Lavishly illustrated with photos from the 19th Century, this hard to find information was first published in 1890's and includes the stories of "California's Rand", details about Chinese miners, how one early miner named Baker struck it rich and also the story of Alphonzo Bowers, who invented the first hydraulic gold dredge. **8.5" X 11", 54 ppgs, Retail Price: $9.99**

More Mining Books

Prospecting and Developing A Small Mine - Topics covered include the classification of varying ores, how to take a proper ore sample, the proper reduction of ore samples, alluvial sampling, how to understand geology as it is applied to prospecting and mining, prospecting procedures, methods of ore treatment, the application of drilling and blasting in a small mine and other topics that the small scale miner will find of benefit. **8.5" X 11", 112 ppgs, Retail Price: $11.99**

Timbering For Small Underground Mines - Topics covered include the selection of caps and posts, the treatment of mine timbers, how to install mine timbers, repairing damaged timbers, use of drift supports, headboards, squeeze sets, ore chute construction, mine cribbing, square set timbering methods, the use of steel and concrete sets and other topics that the small underground miner will find of benefit. This volume also includes twenty eight illustrations depicting the proper construction of mine timbering and support systems that greatly enhance the practical usability of the information contained in this small book. **8.5" X 11", 88 ppgs. Retail Price: $10.99**

Timbering and Mining - A classic mining publication on Hard Rock Mining by W.H. Storms. Unavailable since 1909, this rare publication provides an in depth look at American methods of underground mine timbering and mining methods. Topics include the selection and preservation of mine timbers, drifting and drift sets, driving in running ground, structural steel in mine workings, timbering drifts in gravel mines, timbering methods for driving shafts, positioning drill holes in shafts, timbering stations at shafts, drainage, mining large ore bodies by means of open cuts or by the "Glory Hole" system, stoping out ore in flat or low lying veins, use of the "Caving System", stoping in swelling ground, how to stope out large ore bodies, Square Set timbering on the Comstock and its modifications by California miners, the construction of ore chutes, stoping ore bodies by use of the "Block System", how to work dangerous ground, information on the "Delprat System" of stoping without mine timbers, construction and use of headframes and much more. This volume provides a reference into not only practical methods of mining and timbering that may be employed in narrow vein mining by small miners today, but also rare insights into how mines were being worked at the turn of the 19th Century. **8.5" X 11", 288 ppgs. Retail Price: $24.99**

A Study of Ore Deposits For The Practical Miner - Mining historian Kerby Jackson introduces us to a classic mining publication on ore deposits by J.P. Wallace. First published in 1908, it has been unavailable for over a century. Included are important insights into the properties of minerals and their identification, on the occurrence and origin of gold, on gold alloys, insights into gold bearing sulfides such as pyrites and arsenopyrites, on gold bearing vanadium, gold and silver tellurides, lead and mercury tellurides, on silver ores, platinum and iridium, mercury ores, copper ores, lead ores, zinc ores, iron ores, chromium ores, manganese ores, nickel ores, tin ores, tungsten ores and others. Also included are facts regarding rock forming minerals, their composition and occurrences, on igneous, sedimentary, metamorphic and intrusive rocks, as well as how they are geologically disturbed by dikes, flows and faults, as well as the effects of these geologic actions and why they are important to the miner. Written specifically with the common miner and prospector in mind, the book will help to unlock the earth's hidden wealth for you and is written in a simple and concise language that anyone can understand. **8.5" X 11", 366 ppgs. Retail Price: $24.99**

Mine Drainage - Unavailable since 1896, this rare publication provides an in depth look at American methods of underground mine drainage and mining pump systems. This volume provides a reference into not only practical methods of mining drainage that may be employed in narrow vein mining by small miners today, but also rare insights into how mines were being worked at the turn of the 19th Century. **8.5" X 11", 218 ppgs. Retail Price: $24.99**

Fire Assaying Gold, Silver and Lead Ores - Unavailable since 1907, this important publication was originally published by the Mining and Scientific Press and was designed to introduce miners and prospectors of gold, silver and lead to the art of fire assaying. Topics include the fire assaying of ores and products containing gold, silver and lead; the sampling and preparation of ore for an assay; care of the assay office, assay furnaces; crucibles and scorifiers; assay balances; metallic ores; scorification assays; cupelling; parting' crucible assays, the roasting of ores and more. This classic provides a time honored method of assaying put forward in a clear, concise and easy to understand language that will make it a benefit to even beginners. **8.5" X 11", 96 ppgs. Retail Price: $11.99**

Methods of Mine Timbering - Originally published in 1896, this important publication on mining engineering has not been available for nearly a century. Included are rare insights into historical methods of timbering structural support that were used in underground metal mines during the California that still have a practical application for the small scale hardrock miner of today. **8.5" X 11", 94 ppgs. Retail Price: $10.99**

The Enrichment of Copper Sulfide Ores - First published in 1913, it has been unavailable for over a century. Topics include the definition and types of ore enrichment, the oxidation of copper ores, the precipitation of metallic sulfides. Also included are the results of dozens of lab experiments pertaining to the enrichment of sulfide ores that will be of interest to the practical hard rock mine operator in his efforts to release the metallic bounty from his mine's ore. **8.5" X 11", 92 ppgs. Retail Price: $9.99**

A Study of Magmatic Sulfide Ores - Unavailable since 1914, this rare publication provides an in depth look at magmatic sulfide ores. Some of the topics included are the definition and classification of magmatic ores, descriptions of some magmatic sulfide ore deposits known at the time of publication including copper and nickel bearing pyrrohitic ore bodies, chalcopyrite-bornite deposits, pyritic deposits, magnetite-ileminite deposits, chromite deposits and magmatic iron ore deposits. Also included are details on how to recognize these types of ore deposits while prospecting for valuable hardrock minerals. **8.5" X 11", 138 ppgs. Retail Price: $11.99**

The Cyanide Process of Gold Recovery - Unavailable since 1894 and released under the name "The Cyanide Process: Its Practical Application and Economical Results", this rare publication provides an in depth look at the early use of cyanide leaching for gold recovery from hardrock mine ores. This volume provides a reference into the early development and use of cyanide leaching to recover gold. **8.5" X 11", 162 ppgs. Retail Price: $14.99**

California Gold Milling Practices - Unavailable since 1895 and released under the name "California Gold Practices", this rare publication provides an in depth look at early methods of milling used to reduce gold ores in California during the late 19th century. This volume provides a reference into the early development and use of milling equipment during the earliest years of the California Gold Rush up to the age of the Industrial Revolution. Much of the information still applies today and will be of use to small scale miners engaging in hardrock mining. **8.5" X 11", 104 ppgs. Retail Price: $10.99**

Leaching Gold and Silver Ores With The Plattner and Kiss Processes - Mining historian Kerby Jackson introduces us to a classic mining publication on the evaluation and examination of mines and prospects by C.H. Aaron. First published in 1881, it has been unavailable for over a century and sheds important light on the leaching of gold and silver ores with the Plattner and Kiss processes. **8.5" X 11", 204 ppgs. Retail Price: $15.99**

The Metallurgy of Lead and the Desilverization of Base Bullion - First published in 1896, it has been unavailable for over a century and sheds important light on the the recovery of silver from lead based ores. Some of the topics include the properties of lead and some of its compounds, lead ores such as galenite, anglesite, cerussite and others, the distribution of lead ores throughout the United States and the sampling and assaying of lead ores. Also covered is the metallurgical treatment of lead ores, as well as the desilverization of lead by the Pattinson Process and the Parkes Process. Hofman's text has long been considered one of the most important early works on the recovery of silver from lead based ores. 8.5" X 11", 452 ppgs. Retail Price: $29.99

Ore Sampling For Small Scale Miners - First published in 1916, it has been unavailable for over a century and sheds important light on historic methods of ore sampling in hardrock mines. Topics include how to take correct ore samples and the conditions that affect sampling, such as their subdivision and uniformity. Particular detail is given to methods of hand sampling ore bodies by grab sample, pipe sample and coning, as well as sampling by mechanical methods. Also given are insights into the screening, drying and grinding processes to achieve the most consistent sample results and much more. 8.5" X 11", 124 ppgs. Retail Price: $12.99

The Extraction of Silver, Copper and Tin from Ores - First published in 1896, it has been unavailable for over a century and sheds important light on how historic miners recovered silver, copper and tin from their mining operations. The book is split into three sections, including a discussion on the Lixiviation of Silver Ores, the mining and treatment of copper ores as practiced at Tharsis, Spain and the smelting of tin as it was practiced by metallurgists at Pulo Brani, Singapore. Also included is an overview and analysis of these historic metal recovery methods that will be of benefit to those interested in the extraction of silver, copper and tin from small mines. 8.5" X 11", 118 ppgs. Retail Price: $14.99

The Roasting of Gold and Silver Ores - First published in 1880, it has been unavailable for over a century and sheds important light on how historic miners recovered gold and silver rom their mining operations. Topics include details on the most important silver and free milling gold ores, methods of desulphurization of ores, methods of deoxidation, the chlorination of ores, methods and details on roasting gold and silver ores, notes on furnaces and more. Also included are details on numerous methods of gold and silver recovery, including the Ottokar Hofman's Process, the Patera Process, Kiss Process, Augustin Process, Ziervogel Process and others. 8.5" X 11", 178 ppgs. Retail Price: $19.99

The Examination of Mines and Prospects - First published in 1912, it has been unavailable for over a century and sheds important light on how to examine and evaluate hardrock mines, prospects and lode mining claims. Sections include Mining Examinations, Structural Geology, Structural Features of Ore Deposits, Primary Ores and their Distribution, Types of Primary Ore Deposits, Primary Ore Shoots, The Primary Alteration of Wall Rocks, Alterations by Surface Agencies, Residual Ores and their Distribution, Secondary Ores and Ore Shoots and Vein Outcrops. This hard to find information is a must for those who are interested in owning a mine or who already own a lode mining claim and wish to succeed at quartz mining. 8.5" X 11", 250 ppgs. Retail Price: $19.99

Garnets: Their Mining, Milling and Utilization - First published in 1925, it has been unavailable since those days and sheds important light on the mining, milling and utilization of garnets. Included are details on the characteristics of garnets, where they are found and how they were mined. 78 ppgs, 10.99

Gemstones and Precious Stones of North America - Leading mining historian Kerby Jackson introduces us to a classic mining publication on the gems and precious stones of the United States, Canada and mexico. First published in 1890, it has been unavailable since those days and sheds important light on the gems and precious stones that may be found in North America. Included are chapters on diamonds, corundum, sapphire, ruby, topaz, emerald, disapore, spinel, turquoise, tourmaline, garnets, beyrl, peridot, zircon, quartz crystals, feldspars, pearls and many others. Included are details on where these gems and precious stones may be found throughout North America, as well as their characteristics. 360 ppgs, 24.99

Mining Camps and Mining Districts - First released in 1885 by Charles Howard Shinn under the title "Mining Camps: A Study in American Frontier Government", this publication offers a unique look at how early gold miners established their own forms of representative government during the California Gold Rush. Drawing on the the early mining codes of mideviel German miners in the Harz Mountains, on the mining customs of the Cornish tin miners and early Spanish mining laws introduced into California, the miners established the first governments in the American West. 340 ppgs, 24.99

BLM Field Handbook for Mineral Examiners - Leading mining historian Kerby Jackson introduces us to a classic mining publication on mine evaluation. First published in 1962, this work sheds important light on the techniques of BLM Mineral Examiners to perform validity on mining claims. 132 ppgs, 10.99

Six Months In The Gold Mines During The California Gold Rush - Unavailable since 1850, this important work is a first hand account of one "49'ers" personal experience during the great California Gold Rush, shedding important light on one of the most exciting periods in the history of not only California, but also the world. Compiled from journals written between 1847 and 1849 by E. Gould Buffum, a native of New York, "Six Months In The Gold Mines During The California Gold Rush" offers a rare look into the day to day lives of the people who came to California to work in her gold mines when the state was still a great frontier. **8.5" X 11", 290 ppgs. Retail Price: $19.99**

The Discovery of Gold in Australia - First published in 1852, it has been unavailable since those days and sheds important light on Australia's gold mining history. Included are rare communications between British agents and the British Crown when gold was first discovered in Australia in 1851. This rare text contains hard to find details on Australia's first mining camps and Britain's early attempts to provide for the orderly regulation of gold mines in that part of the world. Also of interest are hard to find extracts of articles that appeared in the early colonial newspapers that did their best to report on Australia's gold rush as it took place.
102 ppgs, 10.99